THROUGH
THE EYE OF THE
STORM

Thank you — for your interest in these wonderful people Linda!

THROUGH
THE EYE OF THE
STORM

Blessings,
Cho

A BOOK DEDICATED
TO REBUILDING WHAT
KATRINA WASHED AWAY

5 - 20 - 06

CHOLENE ESPINOZA

CHELSEA GREEN PUBLISHING COMPANY
WHITE RIVER JUNCTION, VERMONT

All Bible verses are excerpted from the New Revised Standard Version (NRSV).

Editor: John Barstow
Managing Editor: Marcy Brant
Copy Editor: Collette Leonard
Designer: Peter Holm, Sterling Hill Productions
Design Assistant: Daria Hoak, Sterling Hill Productions

Printed in Canada
First printing, March 2006
10 9 8 7 6 5 4 3 2 1

Our Commitment to Green Publishing
Chelsea Green sees publishing as a tool for cultural change and ecological steward-ship. We strive to align our book manufacturing practices with our editorial mission, and to reduce the impact of our business enterprise on the environment. We print our books and catalogs on chlorine-free recycled paper, using soy-based inks, whenever possible. Chelsea Green is a member of the Green Press Initiative (www.green-pressinitiative.org), a nonprofit coalition of publishers, manufacturers, and authors working to protect the world's endangered forests and conserve natural resources.
      Through the Eye of the Storm was printed on Legacy, a 100 percent post-consumer-waste recycled, old-growth-forest-free paper supplied by Webcom.

Library of Congress Cataloging-in-Publication Data

Espinoza, Cholene, 1964–
  Through the eye of the storm : a book dedicated to rebuilding what Katrina washed away / Cholene Espinoza.
    p. cm.
  ISBN-13: 978-1-933392-18-9
  ISBN-10: 1-933392-18-5
  1. Hurricane Katrina, 2005. 2. Disaster relief—Mississippi—Gulf Coast—Case studies. 3. Church work with disaster victims—Mississippi—Gulf Coast—Case studies. 4. Disasters—Mississippi—Gulf Coast. 5. Espinoza, Cholene, 1964– I. Title.
  HV636 2005 .U6 E87 2006
  976'.044–dc22

                          2006004998

Chelsea Green Publishing Company
Post Office Box 428
White River Junction, VT 05001
(800) 639-4099
www.chelseagreen.com

Dedicated to Katrina's survivors:
*Your hope, grace, courage, and love
have liberated my soul.*

CONTENTS

AUTHOR'S NOTE

All of my proceeds from the sale of this book will go to build and support a community/education center that will serve the Katrina survivors of Harrison County on the Gulf Coast of Mississippi. This community center is the first phase in fulfilling the vision Rev. Rosemary Williams has for her community—to create opportunity for children and young adults to learn and develop, to dream and achieve those dreams.

Five acres of land will be developed to provide young adults with GED, computer, and other job training that will give them the skills to participate in the recovery of their community; Katrina washed away many jobs previously held in the area. During non-school hours, the center will serve the children of the community as the only after-school facility in the area. Eventually, routine healthcare education and services will also be provided at the center.

There was one Boys and Girls Club in Pass Christian, but Katrina destroyed it. The nearest facility for children is located in the town of Gulfport, Mississippi, approximately twenty minutes away, and many parents cannot afford to make the trip.

To date the land has been purchased and the plans have been drawn up for the center. It will be located at 6815 Kiln-DeLisle Road in Pass Christian, Mississippi. The property is within walking distance of DeLisle Elementary School, which is now the only school in the area for children in kindergarten through twelfth grade. There are approximately one thousand children currently attending the school, most of them in temporary classrooms.

Rev. Rosemary's parish house is across the street from the center at the new Mt. Zion United Methodist Church, which will be completed in the spring of 2006.The church property includes an additional twenty acres on which to develop affordable housing for Katrina's survivors, another vision that Reverend Williams plans to fulfill.

The community center will be funded by an independent nonprofit 501(c)3 organization that was founded by Rev. Rosemary Williams, Rev. Theodore Williams, Shantrell Nicks, Myrick Nicks, Ellen Ratner and me. It is called the **Pass Christian/DeLisle Community Center, Inc.**

The strength of this community center resides in the volunteers and leadership. The Revs. Rosemary and Theodore Williams, along with many of their congregants, are grandparents and retired teachers. The center will also be outfitted to accommodate volunteers from outside of the local area. There is still an enormous demand for elbow grease as the residents of this community attempt to recover with very few resources. All help is welcome!

CHOLENE ESPINOZA
February 2006

one

STORMS

I have seen many faces of war. I have seen war from a distance, when flying so high I could see the curvature of the earth, yet was still close enough to see the unmistakable flashes of light, indicating that someone is killing and someone else is being killed. I have been on the battlefield—able to see, taste, hear, and feel the death and destruction of war. Flying is the one thing that has allowed me to live with the memories of war. I have always been able to retreat to a place high enough to see a clear horizon, vast enough to recover balance from the disorienting haze of human suffering.

I can think of only one time when flying could not transport me high enough to see the horizon. It was the last week of the summer of 2005, the summer that ended with Hurricane Katrina.

The suffering left in Katrina's wake generated a gravitational pull that ground down my spirit and paralyzed my soul. I could not fly high enough to reach above her path of pain and loss. Her survivors were part of the human family, part of my family, part of my country, and they were living in a hell—a hell I called home.

Even my encounters with war and terror left me unprepared for the totality of destruction I witnessed two weeks after Katrina struck the cities, towns, parishes, and people of the Gulf Coast of Mississippi and Louisiana. Hundred-year-old oak trees pulled from their roots as though they were weeds; a swimming pool standing as the only evidence that a two-hundred-room hotel had ever existed; an orange X spray-painted on a house by a search and rescue team indicating the fate of the previous resident—either dead on the premises or unknown; three homes moved half a block; weary women sifting through discarded old clothing in parking lots that magnified the heat like frying pans; no street signs; no traffic lights; power lines draped across yards.

A dead, silent, deserted American city under military occupation; the smell of raw sewage; a film of hazardous waste covering the neighborhoods of small brick homes; the slowly wagging tails of dogs as they waded through black water and filth in search of food.

Soaked mattresses, ruined appliances, rotten food, and broken sticks of furniture being dragged to the street; the heat magnified by the humidity, cultivating a cancer of black mold that burns the lungs; the vacant stares of adults who were left to survive on the good will of others; the children with bright smiles and sparkling eyes, giggling and playing as if this was the Fourth of July, 2005—the high point of the last summer before Katrina washed away their homes.

I've seen the aftermath of many storms in my life. Some have left a thick, disorienting fog and some have cleared the air for better visibility.

I am a professional airline pilot. Pilots are trained to avoid storms. We fly around them or over them, but never intentionally through them. The closest I have come to flying through a storm was while landing at the Louis Armstrong Airport in New Orleans four days before Hurricane Katrina swept in off the Gulf. We touched down moments before a torrential downpour and lightning closed the airport. We have an expression in aviation: "I'd rather be lucky than good." That day we were lucky.

After waiting around the airport for four hours, we took off again on our third and final flight of the day. As we climbed in altitude, I looked down on New Orleans with a strange feeling of nostalgia—as though I was saying good-bye to an old friend whom I would never see again. I knew there was a hurricane off the coast of Florida, but I had no idea that Katrina would slam into New Orleans the following Sunday.

My premonition was fulfilled the following week as I watched New Orleans drown. How many of my passengers were now trapped in a watery version of Dante's inferno?

I had a similar experience the Friday night before that fateful Tuesday morning of September 11, 2001. I flew a plane, fully loaded with passengers, to my home city of New York. Air Traffic Control cleared us to fly my favorite approach, "to the lady and up the river." We flew to the Statue of Liberty and right by the Twin Towers. As I looked at them, I thought about how fortunate those people were to be able to see their world as I did, from high in the sky.

The following Tuesday I watched television in horror as some of those same people I had flown by (or with) that Friday night chose to jump to their death rather than be consumed by fire.

Next, the news anchor announced that United Airlines Flight 93 was missing. Flight 93 sounded familiar. I checked my flight schedule, found Flight 93's origin and destination, and confirmed what I instinctively knew. I had planned to be a passenger on that flight. I was to fly from Newark, New Jersey, to San Francisco in order to begin my first trip as a new captain from San Francisco on Wednesday, September 12, 2001. My flight from San Francisco had been cancelled; otherwise I would have been a passenger on Flight 93.

I turned away from the television screen and did not tune in again for several weeks. I could not bear to hear the names of my fellow crew members who were killed that day.

New York City was in shock. Grown men and women would suddenly burst into tears while walking down the street. The sidewalks around the city's hospitals were crowded with the families of the missing. Some had infants in strollers or small children who clutched the hands of the adults in tears. The tears were contagious.

I felt helpless. I felt as though I should be doing something. I live less than a couple of miles from Ground Zero, but, understandably, entry was prohibited. This was my city, and these were my people who had died. What remained was a wake of grief so wide that it engulfed the entire world—and there was nothing I could do to help.

I waited for United Airlines to call me to fly again. I was eager to return to the one thing that meant the most to me: flying. I needed the escape and the sense of purpose that flying gives me. Flying demands that I block out emotion, which I was all too eager to do.

Pilots are by nature very compartmentalized. We don't allow the contents of one compartment to spill into the next. Strapping into a jet is like turning on a switch or pushing a circuit breaker. The world, physical and emotional, works as it should as long as there is open sky ahead. The 9/11 attack threatened to overwhelm the walls between compartments. The sanctity of my sky had been violated.

The skies were finally opened again and I returned to the world I understood. My first destination was Boston. The next morning my copilot and I checked into United Airlines Boston Operations. They already had a memorial display of the crew members killed on our planes. A week had passed since 9/11, but this was the first time I could bear to look for our fallen. I was overcome with despair and guilt as I saw the face of a friend I had known for twenty years. LeRoy Homer Jr. was the copilot on Flight 93. I was supposed to have been one of his passengers on that flight.

LeRoy and I were classmates at the Air Force Academy. We graduated together in 1987. When I lived in Manhattan Beach, California, he would meet me on the beach during his layovers. He rollerbladed and I ran. LeRoy, or Lee, as I called him, always had a smile on his face—even as a freshman at the Air Force Academy. He seemed to have an advantage over everyone, even the upperclassmen. Nothing fazed him.

I had just seen Lee on a layover in London during the summer before the world changed. He showed me pictures of his new daughter and beautiful wife, Mclode. I listened to him describe his daughter as we walked through Hyde Park. I remembered him saying, "I'm so happy. Every day is new. They grow so fast. I can't wait to get home to see how she's changed."

A lump began to form in my throat. The emotion that I had been suppressing for over a week was overwhelming me. I went to the bathroom, washed my face, and used a tactic I had developed at the Air Force Academy, the Scarlett O'Hara Strategy: I'll think about it tomorrow.

I put on my captain's hat, walked to the aircraft, and began to perform my preflight checks. The flight attendant came up to the cockpit to tell me we had some family members of the victims of American Airlines Flight 11 on board (the flight that had taken off from Boston and crashed into the north tower of the World Trade Center). I thanked her for telling me. I knew that the right thing to do was to go back there and tell them how sorry I was, but I wasn't strong enough to do it without breaking down.

I was on my way to LaGuardia Airport on September 11, 2005. I again struggled to keep my emotions at bay and my focus on flying. As I drove over the Triboro Bridge, I looked toward midtown Manhattan. Normally this view is inspiring, but the beauty of the city, standing proudly in the morning sunshine, did not lift my sadness. I flashed back to the sorrow of that day four years earlier. This anniversary was worse: another tragedy had affected our nation.

I was drawn to every image, telecast, and newspaper article available on Hurricane Katrina. I read and watched, hoping that a "breaking news" message would stream across the television announcing that our fellow citizens had been rescued. But the nightmare continued. My sleep was littered with dreams of suffering people; I could see the faces of desperate mothers and children. I would wake up in the morning exhausted, feeling as though I had been up all night. I was ashamed and frustrated that I could not help them.

All my frustration and anger bore down on my government. I had been to Iraq twice—once as an embedded journalist with the First Tank Battalion of the United States Marine Corps, and again the following June, after the invasion. We had shipped almost two hundred thousand people and enough firepower to overthrow a government and attempt to rebuild an entire nation more than six thousand miles, but here, within our own borders, we could not provide food, water, protection, or an escape for the victims of Hurricane Katrina.

As I watched government official after official tell his version of the story on television—making excuses or blaming others, including the victims—I was reminded of similar arguments offered by officials when things went horribly wrong in Iraq. They all seemed to say the same thing: "I take responsibility, but it's not my fault," or, "This is an enormous task."

Hurricane Katrina exposed all of our hidden vulnerabilities, inadequacies, and inequities. Like the bright, clear air that rushes in after a great storm, she cast a light into the dark corners that hide the poverty and suffering of "great" nations. No amount of spin or rationalization could change the fact that our nation was ill-equipped to take care of its own, before Katrina and after.

One morning after another sleepless night, I opened my Bible. I hadn't opened my Bible since returning from the war in Iraq. I still carried it in my suitcase on every trip, and kept it next to my bedside at night as though it was a good luck charm, but it had lost its status as my compass for life.

I opened it to the following passage: "Let us therefore no longer pass judgment on one another, but resolve instead never to put a stumbling block or hindrance in the way of another." (Romans 14:13)

This passage cooled my anger, but I was still frustrated.

I flipped through a few more pages and found the parable of the mustard seed. "The kingdom of heaven is like a mustard seed that someone took and sowed in his field; it is the smallest of all the seeds, but when it has grown it is the greatest of shrubs and becomes a tree, so that the birds of the air come and make nests in its branches." (Matthew 13:31)

This parable turned my thoughts toward the tiniest of Katrina's survivors, the children—the mustard seeds—the smallest and most vulnerable. But if we did right by them, we could give them a new opportunity and help them grow to become the leaders and protectors of their communities.

Finally I turned to the words, "So the last will be first, and the first will be last." (Matthew 20:16)

Throughout my life, I have found strength in this passage. It carries a message of hope into worlds filled with injustice and suffering. It says to the vulnerable, weary, and worn, "One day, your pain shall end and you will be the first to be cared for, not the last."

These scriptures penetrated my hardened shell of anger and opened a tiny crack of hope. I wept for the first time since returning from Iraq. My fog of grief began to clear.

two

A LITTLE CHILD SHALL LEAD THEM

Like the unmistakable green and white beacons that allow pilots to find airports amid all the other city lights when flying in darkness or low visibility, the Bible's scriptures caused me to turn toward the light. In retrospect, perhaps it was more a matter of the light shining its beam toward me.

I was challenged by the passages about the mustard seed and the last becoming the first. It was as if I was being pushed to action. I was restless, but unable to act. I was stuck. My brother Chip, a management consultant and ordained minister, had taught me about "being stuck." He bases much of his consulting philosophy on the teachings of Rabbi Edwin Friedman, an expert in family systems theory. In his book *Generation to Generation: Family Process in Church and Synagogue*, Rabbi Freidman talks about the phenomenon of being stuck. Individuals, families, companies, and even countries can become stuck. Symptoms of being stuck include anger, violence, litigiousness, and resignation. People who are stuck can be plagued by feelings of anxiety, uncertainty, anger, frustration, or alienation. They are highly reactive, and they tend to blame others.

I had been experiencing these symptoms on and off since September 11, 2001. It was as if being stuck was some sort of addiction or illness that I could not completely shake. I remember telling Ellen Ratner, my life partner, that before Katrina I finally felt like life was almost normal again. I had physically and psychologically made it through 9/11, covering the Iraq War, and United Airlines filing bankruptcy. "Katrina has changed everything," I told Ellen.

I saw the *here-we-go-again* look on Ellen's face. Ellen and I met in November of 2002. She runs Talk Radio News Service, a Washington, D.C., news bureau that covers the White House, Capitol Hill, the Supreme Court, and the Pentagon. She is also a liberal commentator on the Fox News Channel. Ellen is a 4-foot, 10-inch bundle of energy. I often feel as though I have attached myself to a speedboat as I attempt to "nose surf" behind her.

If she were a dog, she would be a dachshund—an immaculately groomed, fearless little creature that lives to confront larger creatures, even if they happen to be presidents, congressmen, or Syrian government security agents. She is fiercely loyal to her family and friends. She will defend the weak and weary at her own peril.

Ellen offered me the chance to cover the Iraq War as a reporter for Talk Radio News Service three weeks after we met. Before I returned from Iraq, Ellen used to tell others that I was the only person she had ever met who had no negativity; that I led a life full of hope. This was not an entirely accurate characterization, but that is certainly the image I did my best to project. As they say in Alcoholics Anonymous, "Fake it till you make it."

According to Ellen and my family, I came back from the war a changed person—and not for the better. Ellen called my state of mind posttraumatic stress syndrome. (She had been a therapist for more than fourteen years before becoming a journalist.)

Although there were times when the Marines I was with were under attack, I was not afraid to die. I never had nightmares that I was going to die. I told Ellen that her diagnosis was psychobabble and that I hadn't been there long enough to have developed a syndrome. Nevertheless, images of the young troops I had been with in Iraq dominated my dreams for two years. A part of my mind was dedicated to them all the time, as though there was a constant power-draw on my psyche.

While I was trying to maintain an upbeat persona, internally I had all the symptoms of being stuck. I was anxious about my future as a professional pilot in an industry that was hanging on a thread, and I was angry with our government for seemingly driving our military—not just an instrument of war, but young men and women with mothers and fathers and brothers and sisters and husbands and wives and children—into a dangerous and deadly corner with no safe way out.

Each time I saw a news report or read an article about more Iraqi casualties, I thought about the Iraqis I had met. The people of the Middle East are the most generous hosts I have ever encountered. They welcomed Ellen and me into their small homes and honored us as if we were royalty. They fed us what little food they had. They were so proud of their homes, but the chronic uncertainty surrounding everything—security, food, jobs, electricity—had worn them down.

Although intellectually I understood that I was stuck, I was unable to think my way out of it. But then I attended a seminar given by my brother Chip to a group of Santa Fe County employees. My sister, Valerie Espinoza, had just taken office as the new Santa Fe county clerk. Chip celebrated her new position by donating his time to train her staff in organizational change. He offered an insight that clicked for me. "It's easier to act your way into a feeling than to feel your way into an action."

This reminded me of the Aristotelian philosophy I had learned at the Air Force Academy, "To be is to do." The Nike Corporation simplified the same notion with their "Just Do It" advertising campaign. I decided to start acting as though everything was going to be okay. First, I purchased new pilot uniform shirts. I had been wearing the same ones for six years, not wanting to buy new ones in case my airline didn't pull through. I stopped reading about the woes of the airline industry. I stopped complaining about pay cuts and longer work hours. As my language and behavior changed, my mood changed.

It was more difficult, however, to shake my worry over Iraq. Yet as the months went by, I became more and more numb to the pain I witnessed in Iraq and to the frustration I felt. I had started to build up immunity to the violence; it was as though I had taken small amounts of poison until my body finally adopted its own antidote—resignation.

And then Katrina blew in and blew me right off my feet. First, I blamed our commitments in Iraq for the human disaster in the Gulf Coast. I wanted to give a "blanket party" for every government official and pundit I saw on television who had enough hubris to make excuses for this gross dereliction of duty; or worse, to blame the victims themselves.

I learned about blanket parties at the Air Force Academy. The parties were used as a behavior modification tool. They were safer and more effective than turning someone in for inappropriate behavior. The official route to justice was too cumbersome and ran the risk of blowback from the upperclassmen.

Instead, we gave the perpetrator a blanket party. We went into the subject's room in the middle of the night (cadets were not permitted to lock their doors), threw one of our government-issued blue wool blankets over him and proceeded to beat him until we felt better. We then ran like hell back to our rooms before the person knew what, or who, had hit him.

Rabbi Friedman would likely disapprove of blanket parties.

He does not mention violence as an antidote to being stuck. In fact, violence usually results in more violence, which results in being more stuck. Friedman says that the best way to get unstuck is through serendipity or adventure. I have been blessed by a lifetime full of serendipity and adventure. Many of the highlights of my life—attending the Air Force Academy, learning to fly, flying above seventy thousand feet in a U-2, reporting from the Middle East, my relationship with Ellen— all grew out of elements of serendipity and adventure. But somewhere along the way I lost what the French call *joie de vivre*, the joy of living, the zest for life. As I was struggling to renew my love for life, the storm carried me back to a place where everything else in life seemed trivial and worthless in light of the magnitude of the event.

I wanted to go to the Gulf Coast as soon as possible. I felt as though the hopeful scriptures were directing me to the region. I was also drawn to the coast for the same reason I went back to Iraq for the reconstruction: I did not want to sit and criticize the actions of others without seeing it for myself. Unlike in Iraq, this time I had an opportunity not only to understand and report on what was going on, but also to help—even if it was in a small, barely measurable way.

Ellen and I had planned to take a vacation in the middle of September. This was to be a real vacation. Our other "vacations" had been spent in Iraq, Syria, Jordan, the Palestinian territories, and other such garden spots. We used to joke that we needed a new travel agent. This vacation was going to be different. We were going to stay in the United States. Ellen, the consummate planner, had mapped out a trip that included Yellowstone, Jackson Hole, and my home state of New Mexico—the "Land of Enchantment."

Then after another sleepless night rife with Katrina dreams I announced to Ellen that we couldn't go on vacation as planned. This was one week after Katrina struck. I did not have any ideas about how we could help Katrina's survivors, but I was not in the mood to vacation. She didn't argue, but she said, "Okay, but you can't just sit around and be depressed about it."

Ellen gave me three choices:

1. She could go on vacation without me.
2. I could write some articles about my thoughts on Katrina as they related to the state of our nation.
3. We could go to the Gulf Coast and volunteer to help.

I thought about Ellen's three choices for about half a second and said, "Let's go!" Ellen had that look she gets when she's about to do something she knows is a little crazy. It's the same look my four-year-old-niece, Charli, gets when she dives off the backyard wall into the swimming pool.

"I'll call that woman I met on the plane!" Ellen announced.

The previous week Ellen had called me from a plane. She had spent a few days in Odessa, Texas, working on her book *Ready, Set, Talk!* with my mother, an editor, and she was on her way back to Washington, D.C. She was on the final leg of the trip, waiting for the passengers in Jackson, Mississippi, to board.

She stopped talking to me mid-sentence and said, "Oh, no."

"What's the matter? Do you have a mechanical problem?"

"Worse."

"Worse? You're cancelled?"

"There's this woman coming down the aisle with two kids and they are heading right for my row," Ellen replied in a pained voice.

I heard Ellen saying hello, playing nicey-nice with the

woman she didn't want sitting next to her. Then, in a voice she only uses on children and congressmen, I heard her say, "How are YOU???"

"They're crawling all over me," she said under her breath to me, through what must have been clenched teeth from a fake smile.

She said to her seatmates, "Oh no, that's FINE. They're fine, *really*." Back to me, "I gotta go!"

I'm used to having Ellen end our conversations abruptly. She drops in to nineteen radio shows each weekday, from 6:45 AM until 8:30 PM, and some on the weekends, bantering with mostly conservative talk show hosts. Ellen pushes everything in life to the last minute, even conversations. So when it's time to get off the phone, it really is time to get off the phone.

She landed a few hours later and called to debrief me on her trip. I assumed she would have the woman's whole life story. Ellen has spent the last thirty years of her professional life listening and talking—half the time as a therapist and half the time as a journalist and talk show host. Ellen has an innate desire to know everything there is to know about everyone. I, on the other hand, take the military secrets approach and assume that I don't need to know every detail of anyone's life, let alone that of a complete stranger.

"How was the flight?" I asked.

"The little monsters were crawling all over me the whole flight. I must have read their books to them ten times!"

Although Ellen was illuminating the negative, as she is prone to do, there was an upbeat tone in her voice, as though she had actually enjoyed herself. I could tell that "the little monsters" had made quite an impression.

"What's the story?" I asked, as in, what is this woman's life story? I knew that Ellen had it.

"Yep," she opened. "They were getting away from the hurricane." Then she told me, in her urgent, rapid-fire way, about this incredible woman and her two children.

"The woman's a lawyer. She's black. Her name is Shantrell Nicks. She has two kids, a girl, Taylor—she's three—and a boy, Mason—he's eighteen months. She is married to a vice principal of a huge local high school. Their house had damage from the hurricane, and her husband is working like crazy with the rest of the staff to clean up the high school. She's going to stay with her sister-in-law who lives in Washington, D.C. I'm having dinner with her this week. I'm supposed to call her.

"It sounds like they really need help—especially her aunt, who is a reverend at a small church. It's a mess. I'm going to call her tomorrow. Oh yeah, her mother's a lawyer and part-time judge. How about that?"

Ellen's own private relief effort began the next day when she called Shantrell. She also called Shantrell's aunt, Rev. Rosemary Williams of the Mt. Zion United Methodist Church of DeLisle, Mississippi. She put Rev. Rosemary on air and on the Talk Radio News website. It was a few days later that we decided to scrap our vacation trip and go to the Gulf.

"Rev. Williams, this is Ellen, from Washington, D.C. We would like to come down there. Can you still use our help?"

And so the adventure began.

three

PETAH TIKVA—AN OPENING OF HOPE

I was skeptical when Ellen first suggested that we—two gay women—should drive down to the heart of the Bible Belt, to one of the reddest of the red states, and camp out with two churches. "We're the gay version of the movie, *Guess Who's Coming to Dinner,*" Ellen said to reassure me.

My skepticism was rooted in my past church experiences. While growing up, I had attended the Assemblies of God Church. My mother became a born-again Christian when I was six years old, and going to church was the focal point of my childhood for ten years.

I participated in youth groups, Bible studies, church theater, and summer camp, but I never felt that I belonged. Part of this was because of my family life and part of it had to do with me. We were a blended family when there were not many blended families in church. My mother brought my brother and me to the marriage and my stepfather brought his three children— his daughter Laron and two sons, Phillip and Jeffrey. Some of the time, Laron and Jeffrey lived with their mother. When I was nine years old, *yours, mine, and ours* was completed with the addition of my little brother Jason.

My father's side of the family was also blended. He had my sister Valerie before he married my mother. My brother Chip and I spent summers with them in Espanola, New Mexico.

While I did not fully understand it as a child, our family faced a glass ceiling of sorts because my parents had been divorced and remarried. To remarry is considered adultery, which is a sin. Our church encouraged us to love the sinner, but not the sin. This meant that my stepfather could not be a deacon in the church. My mother could not be one either, for the additional reason that she is a woman.

Although never spoken about or even acknowledged, I sensed that there was always an undercurrent of shame when it came to our church life. My parents had to be married by a justice of the peace because their marriage ceremony was not permitted in the house of God. We were not *pure* and we never would be. But we tried to be pure. My mother had a drama ministry; she wrote, directed, and performed Christian drama. My stepfather was equally committed. They tithed faithfully and gave additional offerings for church building projects and other needs. We sacrificed. We went to every service that was offered, Sunday morning, Sunday night, Wednesday night, and Friday night youth. We tried to be worthy, accepted, and equal.

My father had a similar shame experience with his church, the Catholic Church. His response however, was very different from my mother's and stepfather's. He did not try to be worthy of his church; he simply rejected it.

I very much wanted my father to go to church. I was afraid that if he didn't, he would go to hell. I spent my summers and vacations with him trying to persuade him to go to Mass. He always had the same argument: "My church has rejected me because your mother and I are divorced. Who needs them?" I finally told him one day that if he didn't go to church he would go to hell. He said, "Well, I'd rather be in hell with the sinners than in heaven with the saints." We never spoke about his faith again.

Shame has reverberated in my father's family for more than five hundred years. We are what is known in northern New Mexico as the Crypto, or Hidden Jews. I discovered this last year as I was reading *The Martyr*, a book by Rabbi Martin Cohen. It is the story of the Carvajal family. The Carvajals fled Spain and Portugal following the forced conversions of the 1490s. They sailed to the New World, Mexico, for safety and opportunity. Some members achieved great success—perhaps too much success. The Inquisition again made its way to the Carvajals.

One member of the family, Luis Carvajal, was the governor of Nuevo Leon. He was a true devotee of Christianity, but that was not enough. The Inquisitors imprisoned him and he lost everything while attempting to appeal his conviction, including his life.

The other prominent member of the Carvajal family was the governor's nephew, also named Luis. The younger Luis was as fervent a believer in Judaism as his uncle was in Christianity. He remained committed to Judaism even after being imprisoned for his beliefs. Ultimately, he was burned at the stake.

Many of my ancestors survived the Mexican Inquisition by fleeing to northern New Mexico for safety. Today, the family's practice of Judaism is limited to a few rituals, such as the kosher killing of animals and lighting the menorah.

While reading *The Martyr*, I discovered that I was related to the Carvajals. Dr. Stan Hordes, a prominent historian of the Hidden Jews, had linked my family to the Carvajals. Suddenly I understood the term *genetic memory*, meaning that certain traits, behaviors, and, in some cases, tools for survival are passed from generation to generation. My family passed on the cloak of secrecy and shame created by an Inquisition that forced them into hiding for centuries.

I had a personal shame to add to my family shame: I have known that I was gay since I was a small child. My family could not hide our blended family, but I could hide my being gay, and so I did until I was thirty-eight years old.

I was quite certain that I would go to hell for being gay, and that if Christ returned before I died, I would be "left behind." One year my mother wrote a play about the Rapture. I played the teenager who was left behind. It was my best performance, according to the audience and my mother. They didn't know that my art was imitating my reality.

By the time I was ready to graduate from high school, I could not wait to get out of the Assemblies of God Church. Perhaps it was a natural rebellion. Some teenagers do drugs or drink; I switched churches.

I started attending Catholic Mass as soon as I entered the Air Force Academy. I went to Mass every day my freshman year. It comforted me. At Mass, cadets shared emotions they could not share elsewhere. It was the one place where I could be vulnerable and ask for help. I loved the construct of the service—always knowing what to expect. There is a pacifying rhythm to the Mass. While the Catholic Church shared many of the same social views as the Assemblies of God, it was less overt about them. They had developed a "don't ask, don't tell" mentality long before this notion was introduced into the military by the Clinton administration.

Although the topic was not brought up, let alone talked about, I was still troubled by being gay. One morning, while sitting in the Catholic Chapel of the Air Force Academy, a question popped into my head: "If there were no heaven or hell, would you still be a Christian?" My answer was yes. I believe this question emerged from my inability to reconcile being gay with being a Christian.

I prayed that God would heal me of being gay. Since childhood, being gay had been the thorn in my flesh, a constant irritant to my spirit. But the prayers did not work. This was a problem for me because as far as I was concerned, being gay ranked right up there with the deadly sins. Sin separated me from God and separation from God was its own hell.

The question, "If there is no heaven or hell . . ." forced me to look beyond God as an element of reward or punishment. As I delved into the gospels of Christ, I found a way to live that comforted and empowered me. I focused on loving my neighbor by serving my neighbor—being helpful. I refused to pass judgment on others. I focused on what was good. I lived with the joy of Christ. Most important, I began to focus on others and not on my own personal pain as I struggled to get through my freshman year of harassment and humiliation.

Looking back, I realize that the harassment and humiliation were by institutional design, not the whims of the pubescent powermongers we called upperclassmen. The treatment drove us to depend on and give to one another. Freshman year could not be survived alone. Our favorite sayings were: "The chain is only as strong as the weakest link," and "There's no *I* in *TEAM*."

During one midday training session, the sophomores in my squadron herded our class, freshmen, into a windowless room and turned off the lights. We had no idea what would happen to us next; the military loves the element of surprise. Suddenly I sensed that someone was in front of me. I could feel another person breathing in my face. And then the yelling started: thirty upperclassmen were yelling at the top of their lungs face-to-face with thirty freshmen.

By this time, I'd become a very thick-skinned freshman, yet this terrified me. I felt I would suffocate, but then I felt the warm hand of my classmate, Sean Murphy, make a cup over my clenched fist as I stood at attention. I was not alone, and because I was not alone, I could endure anything. Sean taught me the value of comforting another person even in a time of personal pain. He was killed five years later when his F-15 ejection seat malfunctioned.

Eventually I stopped asking God to heal me of being gay. I tried to ignore the entire matter. In the Air Force, introspection was considered a distraction, a liability, a luxury. And then I fell in love with flying. It has been said that flying is freedom in space and sailing is freedom in time. Airplanes don't care about your race, creed, gender, or sexual orientation. Either you have the skills to fly or you do not. Airplanes are the great equalizers. Most important, airplanes took me to a place where I could view the world from a safe distance, where I could take in its beauty without its pain. I was never lonely in the air. Flying was a gift.

So I focused on flying. Being gay was not at the forefront of my mind as a defect in need of fixing. Even so, I lived for the next eighteen years with a sort of low-grade fever of self-loathing.

After graduating from the Air Force Academy in 1987, my attendance to the Catholic Church dropped off to a paltry few times a year. I took my Grandma Trinnie Espinoza to Mass when I visited her in Espanola, New Mexico, and I sat in Saint Patrick's Cathedral in New York City when I felt alone or depressed.

I attempted to remain centered by reading the Bible as well as several Christian and Jewish authors. I strived to live life as a Christian, which for me meant emulating the life and teachings of Christ. Although being raised in the Assemblies of God Church had its pain and shame, I had received the message of God's love there and did not want to give up on it.

In July of 2002, I had been agonizing over whether to go back into the military after the terrorist attacks of 9/11. I felt compelled to serve again, but I did not think I could go back into hiding under the "don't ask, don't tell" policy, which translates in plain language to: you can be gay, but you will be dismissed for overt statements and acts, or for marriage.

While willing to live a celibate life, I knew that hiding would take its toll on my integrity, as it had before. I wanted to be true to who I was, but I felt a duty to serve my country in uniform again. Inner conflict doesn't agree with me. I was falling into a minor case of self-loathing when I heard a voice that I recognized as God's saying, *"I chose you. I created you in my perfect love. It's time to live up to your responsibility as my creation."* For the first time in my life I was certain that I was a child of God—not an outcast, and not defective.

While I remain convinced that I am a child of God created in perfect love, others still believe that I am an abomination. Some church leaders go as far as to blame me, and others like me, for the mass murders of 9/11 and the destruction of New Orleans. They claim that God punished those cities because they have large gay populations. Most Christians believe, and I agree, that this is a radical, fringe element of the church. Unfortunately, that radical element has gotten a tremendous amount of airtime in recent years, and used this airtime to shape public opinion and national policy. They manipulate our leadership with promises of something even more powerful than campaign contributions—votes.

Religious hardliners drove me further from the church with their support of war. From my perspective it looked as if a deal had been cut between the government—which needed support for war—and the hardliners who needed support for their social agenda. I could find no other explanation. Why would men and women who profess to believe in the love of Christ advocate a policy that even went beyond an eye for an eye? The prevailing policy can be summed up as "kill them before they kill you, and if you kill some innocents in the process, so be it."

Before long, I stereotyped the church—all churches—into that convenient term "the religious right." This was immature and unfair, and embraced the very close-mindedness that I accused the religious hardliners of having. Yet I held this view until I met Rev. Rosemary Williams and her husband, Rev. Theodore Williams.

It was about ten days after Katrina hit when I first spoke to Rev. Rosemary. I called her to learn what people needed and encouraged her not to hold back. Large and small, she itemized each need, from bleach to insurance deductibles, cars to underwear.

Listening to Rev. Rosemary, I began to comprehend the magnitude of loss. Her congregation and community did not even have enough underwear, let alone the resources to rebuild. Rev. Rosemary had been instantly propelled to the position of first and last responder: she was feeding, clothing, and attempting to shelter not only her congregants, but also the greater community. I could hear the fatigue in her voice. At the end of the conversation she said, "We are grateful for anything you bring."

There is a city in Israel named Petah Tikva. *Petah tikva* means "the opening of hope" in Hebrew. That first conversation with Rev. Rosemary was my *petah tikva*. She spoke, I listened. The language she spoke was not only one of need, but also one of opportunity. She opened the door for me to give and, in doing so, to come home, home to my roots—both Christian and Jewish—and home to a community of love, not shame.

four
LOGISTICS

No mission can be completed without logistical support, a supply chain or "log train," as we call it in the military. The engine that drives this train is a "supply troop"—the go-to guy or girl. Being the supply troop is not glamorous. No one knows this person exists unless something goes wrong. It is a thankless, vital job.

I learned the importance of logistical support as a U-2 pilot in Saudi Arabia in December of 1994. The U-2 flies at altitudes above seventy thousand feet, and therefore has a very particular appetite when it comes to fuel. There was plenty of fuel in Saudi Arabia, and it was even cheaper than water, but there was no U-2 fuel. It had to be shipped in and cleared through Saudi customs. The Saudi customs gate is like the gate to heaven described in the book of Matthew: "For the gate is narrow and the road is hard that leads to life and there are few who find it." (Matthew 7:13)

Despite the best efforts of some fine Air Force supply personnel, no one had managed to find a way through the Saudi customs gate. Perhaps we had enough fuel for two missions, but that was it. Dozens of requests to the head supply office met the same response: "Should be any day now." Having nothing better to do as an American woman in Saudi Arabia, I spent a day looking for the person who was *really* in charge of supply. I knew this person would have no impressive title. We had already been through all the personnel with important sounding titles. I was looking for the equivalent of Radar

O'Reilly, the ingenious go-to guy on the television series *M.A.S.H.*

Sure enough, he was an American sergeant buried somewhere over at Saudi customs. At first the sergeant gave me the pat answer: "Should be any day. We don't have any control over the Saudis."

"Sergeant," I said calmly, "I've got your name here and I will have no choice but to include it on Wednesday's mission cancellation message as the person who could not supply the fuel. Do you suppose the joint chiefs will understand how the U-2 squadron could possibly run out of gas in a country with the largest oil reserves in the world?"

"I get it," he said.

Within a day, we were flush with fuel.

Ellen and I did not want to waste time driving a U-Haul 1,300 miles from New York City to Mississippi, so the closer the supply troop and the supplies to the theater of operations, the better.

After Ellen's father died of heart disease when she was ten years old, her mother remarried a man from Memphis, Tennessee. Ellen and her mother moved from Cleveland, Ohio, to Memphis, where she met Eleanor Goodman (now Gipson) at the Lausanne School for Girls in 1965, when they were thirteen years old. As soon as we decided to go to Mississippi, even before we spoke with Rev. Rosemary Williams, Ellen said, "I'll call Eleanor."

Eleanor still lives in Memphis just a few blocks from the home where she grew up, amid a grove of lazy, elegant dogwoods and magnolias that are at least twice her age now. Eleanor is a titanium magnolia. Titanium is lighter than steel. It is resistant to corrosion and has a high strength to weight ratio.

It is capable of wielding more relative force then steel, particularly when directed at an obstruction.

Eleanor wears an apron when she makes coffee. She and her husband, Andy, compete in the Masters' Division of the USA Track and Field competition. Both look at least a decade younger than they claim to be (over fifty years old). Eleanor is nationally ranked in the high jump and Andy is a sprinter. Andy has the demeanor of other sprinters I've known—focused, confident, and quiet, yet he calls complete strangers "baby doll," as in, "Baby doll, could I get another glass of tea?"

Ellen announced to me, "I'm sure Eleanor is awake. I'll call her now." It was 7:30 AM in Memphis.

"Eleanor! Hi! Oh, did I wake you? Are you okay? Surgery? Wisdom teeth? Wow! Infected? Okay, yeah. We want you to come to Mississippi with us. We're going down there to help out with the hurricane. Okay. Call me when you wake up."

Ellen had awakened Eleanor from a pleasant Vicodin fog. She called Eleanor back a few hours later. "What did we talk about this morning?" Eleanor asked.

"Okay, we want to rent a U-Haul truck and load it up with stuff to drive down to this church in Mississippi." Ellen launched into a rapid-fire litany of needs. "We'll meet you in Memphis on Tuesday night. What Tuesday night? Eight days from now. Do you think we can get a U-Haul in Memphis? There aren't any in Jackson. We already checked."

The morning we called Eleanor, her cheek was swollen like a lopsided chipmunk. She was in the middle of trying to sell their home in order to downsize, and was inundated with prospective buyers and constant housecleaning duties. Nonetheless, she pushed her pain and home concerns aside and accepted the position of supply officer.

Eleanor and Andy (who was not personally recruited but

volunteered to give up a week of his life for the procurement segment of the mission) began filling a seventeen-foot U-Haul van with supplies. After my initial conversation with Rev. Rosemary, Eleanor took over and worked directly with Rev. Rosemary and Shantrell (the woman Ellen had met on the airplane) to fill their respective shopping lists—lists that revealed the true depth of devastation in the community. The needs represented the very basics of existence—toilet paper, diapers, canned and dry foods, boxes of juice, soap, bleach, and underwear. Shantrell went to Forest Heights, a housing development in Gulfport, Mississippi, and collected lists of specific needs.

Eleanor, Andy, and Eleanor's sisters spent seven days filling the lists and making up personalized bags for the survivors. They were inundated with donations and offers to help as the word spread that they were going to Mississippi.

Some didn't exactly offer to help; they just announced they were comin' over. One morning Eleanor's friend Kathy called. "You don't have time to make your bed. I'm comin' over now to make it." Kathy knew that Eleanor was showing their house that day and that a properly made bed would enhance the ambience of their beautiful home. Kathy didn't merely offer to help, she jumped in and started helping. If Kathy had asked, "Is there anything I can do for you?" I doubt Eleanor would have said, "Yes, I need you to come and clean my house and make my bed so that it looks like something out of *Southern Living* magazine."

I met another of Eleanor's generous neighbors as we closed up the U-Haul to depart. Mike Palazola owns a produce company, and he donated several cartons of fresh fruit. He came by in his bathrobe to make sure that we had received his shipment. "Y'all have a safe trip. It's great you're goin' down there."

We thanked him for his donation. "Ah, it's really nothin'. I'm just grateful to help."

A few days later Rev. Rosemary stopped us as we were leaving Mt. Zion and said, "Did you get any of those oranges? They are a real treat. Take some. They are the sweetest I've ever tasted." Fresh fruit was a luxury to Katrina's survivors. They had subsisted on a diet of Meals Ready to Eat (MREs) and an occasional hot meal served by volunteers. It's no wonder those oranges tasted so good.

This level of gratitude was overwhelming. We witnessed first-hand the impact that Eleanor and her community had made on the lives of their neighbors to the south. Eleanor was not convinced that we could post a "Mission Accomplished" banner, but without her logistical genius, we never could have taken the hardest step—the first one.

five

OLD FRIENDS

A friend is one who knows all about you and loves
you just the same.

—*Harry Ratner*

As we pulled out of Eleanor's driveway, she asked Ellen if she
wanted to swing by Ellen's old house for memory's sake. Ellen
hesitated slightly and then said, "Su-re," as if she wanted to
swallow the word quickly and skip a trip down bad memory lane.

Ellen does not have fond memories of her time as an adoles-
cent in the South. She arrived, or perhaps I should say landed,
in Memphis in 1965, and it was a hard landing. For starters,
Ellen did not blend in with the genteel and demure young
ladies of the South. Ellen has always been a force to reckon
with. The best illustration of her will as a young girl was when
Ellen's mother, Ann, chose a dress for her to wear for a family
event. Five-year-old Ellen refused to put the dress on. I'm told
that Ellen's willful nature comes from her mother, so, not to be
out-willed by a five-year-old, her mother said, "Fine," and left
Ellen behind—or so she thought. No one leaves Ellen behind.
She emptied her piggy bank, called a cab, and showed up at
Aunt Betty's house in the dress she had chosen for herself.

The move to Memphis isolated thirteen-year-old Ellen from
the people she loved, the cousins and aunts and uncles in
Ohio—the Ratners. Her life centered on this extended family,
which numbered over sixty at that time. The entire family

spent every Saturday night together in celebration of the end of the Jewish Sabbath. Every year from Memorial Day to Labor Day, they lived together in cottages on Lake Erie. There were no boundaries between family members. Their life was one enormous and movable feast of love, food, and fun.

Family, charity, education, and social justice were the cornerstones of Ellen's upbringing. The Ratners' values emerged from their experiences, mostly bad. Ellen's father, Harry, grew up in Bialystok, Russia (now Poland). His family history bears the scars of the brutal Pogroms of Russia. The Pogroms were a form of government-sanctioned rioting against the Jewish people in the eighteenth and nineteenth centuries. They began in Tsarist Russia and spread throughout Eastern Europe. The rioters destroyed homes, sexually assaulted women, and killed thousands of Jews across Russia.

The Pogroms are said to have led to the first immigration of Jews to America. Harry's brother Charlie was the first member of his family to immigrate to America. He joined the U.S. Army and in 1920, after World War I, he had saved enough money to bring the rest of his ten immediate family members to the United States. They settled in Cleveland.

The Ratners were as devoted to their new country as they were to each other. America offered them not only safety, but opportunity as well. They started a small dairy (similar to a convenience store today) and eventually grew an enormously successful building business that is still thriving today.

Their success in America was overshadowed by what was happening in Europe in the late 1930s and 1940s. Some of the extended Ratner family still living in Eastern Europe became victims of the Holocaust, one of the most pervasive manifestations of hate in the history of mankind—the annihilation of approximately six million people.

Ellen's cousin Ida was included in that number. She wrote a song while in the concentration camp titled "Spring is Here Again": "Springtime comes again and with it freedom. When freedom's clock will strike the hour, we will be reborn in freedom. Then, as one, we will re-emerge." Ida died on the way to the hospital after liberation.

I did not fathom the magnitude of this loss until I visited Yad Vashem, the Holocaust memorial in Israel. One large dark room is filled with thousands of lights, each symbolizing the life of a child that was killed. Each light reflects off mirrors that multiply them tens of thousands of times, symbolizing the lives of the children the young victims would have brought into our world had they survived.

The Ratners in America understood and internalized this loss. They took it upon themselves to sponsor and support *anyone* who survived the Holocaust and wanted to make a home in Cleveland. They found shelter for them, gave them work, and set them up in new businesses.

Forty years after the Holocaust, one of these survivors, Edith Simon, described how much Ellen's father had meant to her. With tears in her eyes she said, "When I moved into my new home, I could not believe it. I said, 'What am I going to do? Oh my. I am so grateful but I am so alone. I have no family.' And then Harry said to me, 'I am your new family!'"

To this day Ellen cannot talk about, read, or watch anything pertaining to the Holocaust. She saw and heard all she could bear as a child watching her parents attempt to heal the wounds of its survivors. Ellen recounts an incident when she was in the fifth grade. It was past dinnertime and she had not eaten. She complained to her mother, who was on the phone with a Holocaust survivor. Her mother proceeded to lecture Ellen that these people were more important than anything else in the world, and certainly more important than her dinner.

That moment instilled a lesson into Ellen's mind and heart that guides her every action. When Ellen sees someone in pain, she cannot pass on by; when she sees an injustice, she cannot allow it to continue. Ellen does not have the ability to rationalize pain or injustice. She cannot say, "Oh, it's really not that bad for them. They'll get by. The government is supposed to take care of them. This isn't my responsibility." She does not use the feint of apparent concern to dodge taking concrete action, as in, "It's a shame. I can't believe this is happening."

Ellen felt as if she had been kidnapped and taken to an alien planet when she walked into the Lausanne School for Girls on that first day. She was alone in a strange land with strange customs. The first event was a Mother's Club meeting. The speaker moved to the front of the room and opened with this attention-getter: "Now, girls! You *must not* wear stones before five!" It was several days before Ellen learned that "stones" meant precious stones, as in high-value jewelry.

Soon Ellen began to ensure that everyone at the Lausanne School for Girls knew that she was Jewish, that discrimination was wrong, and that black people were entitled to every right that white people were. Ellen was not well received. Her penchant for justice was a problematic character trait—especially for a thirteen-year-old transplant to the heart of the South at the height of the civil rights movement. Many of her classmates hated her. She still has the Lausanne yearbook with numerous entries, all variations on the same theme. Here are a few: "You're Martin Luther King's First Cousin." "You Nigger lover!" "The South shall rise again!"

Not every girl loathed Ellen. She had one friend, Eleanor Goodman, the daughter of William Wolf Goodman, a renowned attorney. He was one of the first Jews admitted to Harvard Law School; he graduated in 1922. Eleanor's family did not practice Judaism as a religion. Her mother was a nonpracticing Episcopalian. Nonetheless, Eleanor attended the Episcopal Church, and she shared Ellen's values of social justice and charity. Even though Eleanor and Ellen may not talk with each other more than a few times a year, when they do reconnect it is as if they have chatted every day. Their lasting friendship was born out of the trauma of living in the South in the 1960s.

The time I spent with the military (both in uniform and out), from the Air Force Academy to the war in Iraq, taught me that trauma and pain can create the perfect incubator for friendship. Perhaps it is because trauma and pain cause us to lose our grasp and sense of control. It requires us to reach out to each other in mutual trust. My mother calls this "trauma-bonding."

Ellen and Eleanor's friendship reminded me of my friend Scott Wolfe. Scott is also from the South. Looking back, I believe Scott may have saved me during my senior year at the Air Force Academy. He stood by me when doing so ran the risk of collateral damage.

My military record prior to my senior year was one of a so-called model cadet. Cadets are graded academically and militarily on a 4.0 scale. My military grade point average was 3.9 going into the second semester of my senior year. To this day, I cannot put my finger on exactly what happened, but I dug a hole for myself and kept digging. I got into a classic pissing contest with the commanding officer of my squadron. I'll call him Major Jim.

It all started when I embarrassed him in front of a group of officers and their wives, who were having a special visitor's dinner out in the field during Basic Cadet Training (BCT—affectionately known as "BEAST"). I was responsible for training a dozen new Basic Trainees during the summer prior to my senior year at the academy. One evening my "Basics," as we called them, were separated from me for the evening meal.

Another upperclassman sat with them and proceeded to harass them throughout the entire meal. Harassment was standard operating procedure, but normally even the most hardened of hearts would allow the Basics to eat during the last five minutes of the meal. Not so this time. Their plates were still full when they were dismissed from the table.

I called one of them over to my table and put a jar of peanut butter in his field jacket pocket. His eyes sparkled with delight, like a dog being given a treat. It was against policy to take food back to the tents. I said, tongue in cheek, "It's not for you guys. It's for General George. He needs to eat." General George was our element's mascot. He was a rather rotund teddy bear the element had outfitted in a uniform made from an army-green laundry bag.

My eyes followed the Basic as he walked back to his tent. I sensed danger ahead. He was heading right by the officer's mess tent where they were having a family night with their wives. Suddenly an officer (a real officer, not a cadet) stopped him. I knew this officer was a total flame-thrower. His nickname was Major (as in the military rank of major) Asshole.

Most officers at the Academy did not use the immature and unimaginative leadership tools of fear, sarcasm, and ridicule, but this man did. I hoped he would let my Basic pass because they were in the company of civilians (their own wives). Perhaps this is precisely why he stopped the Basic—to put on a little show for the ladies. Then I noticed that my new commanding officer, Major Jim, was also present.

"Mister! What do you have in your pocket?" So much for hoping, I thought, and started making my way over there, as I knew my Basic was going to need some air support. The Basic pulled out the jar of Skippy. "Sir, it is a jar of peanut butter!"

"Do you know it's against the rules to take food from the chow hall?"

"Yes, sir!"

"Then why did you take it?"

The Basic was too afraid to speak.

"WHO authorized you to take it?"

I was coming in from behind. The Basic was petrified. The officer moved in for the kill and got within an inch of the Basic's face.

"Sir. I don't remember," the Basic blurted out.

"Mister! Do you know what an honor violation is?" The Basics had not yet taken the oath of honor: "I will not lie, cheat, or steal, nor tolerate among us anyone who does." The oath is not administered until the end of Basic Training. Nevertheless, the officer's question immediately rolled my Basic. I braced for my name to be announced.

"Sir, it was Cadet Espinoza."

"Sir, I'm Cadet Espinoza."

Major Jim was standing right next to Major Asshole. I saw the look of humiliation and anger on his face. My willful violation of stated policy reflected poorly on Major Jim's leadership. I had managed to humiliate the novice commander in front of the other officers and their wives. Major Jim ordered me to report to him the next day.

He punished me with "40/40/2," forty demerits, forty tours, and two months of restriction. Demerits are like points on your driver's record—only in this case if you get enough of them you are driven out the back gate of the academy for good. Tours are a disciplinary tool that requires the subject to march back and forth with an M-1 rifle in a small courtyard of concrete and embedded pebbles for fifty minutes with a ten-minute break in between tours. The Skippy incident was egregious enough to require me to do forty of these tours. Two months of restriction meant that I could not leave the academy for two months, and I would be required to sign in hourly on weekends and after classes.

I had a strong hint that things were not going to go well for me when Major Jim refused to allow me to be a flight commander in my squadron. In fact, he refused every job for which I was nominated. I was a plain-stripped senior: rank, no status. I lived up to my low position. I adopted the attitude that he was not going to break me.

Major Jim inspected my room daily and I gave him plenty of things to write up. I had trash in my trashcan—a no-no; I had water in my sink—another no-no. We got into a war of one-upmanship, but I intentionally kept my demerit count low with petty crimes in order to avoid being shown the gate.

I had forgotten that a cadet could be thrown out for aptitude as well as for misconduct. Aptitude is subjective; conduct is objective. Major Jim was convinced that I did not have the aptitude to be an Air Force officer. He tried various tactics, like no-notice drug testing, perhaps in hopes that he would find that illegal drug use was the root of my unexplainable behavioral problems in light of my previously good record. He would have me awakened at the crack of dawn and driven to the Cadet Hospital, where I was ordered to stand over a small paper cup in the middle of a room and pee on command. I was clean.

Scott Wolfe was my trusted ally throughout this war. Scott was a recruited quarterback from Picayune, Mississippi. He was one class behind me and lived in my squadron. He used to sneak me off base to take me for a drive and help me feel somewhat normal. I had fewer liberties than the freshmen did during my last quarter at the academy.

Though Scott tried his best to cheer me up, I felt beaten— like a total loser. I didn't care about anything. I'd worked so hard to graduate, but it looked like I might get thrown out first.

I lost the ability to block my emotions, and I started dwelling on my father, who had died from cirrhosis of the liver during winter final exams of my sophomore year. I dwelt on all the promises I had made to him as I knelt by his casket the night of his Rosary. I had promised him that I would show the world, through my life, what he could have done had he been able to quit drinking. These promises had been the source of my motivation for two and a half years, and had sustained me through the grief of my loss. Now they were the source of my shame. My religious beliefs prevented me from taking my own life, but my state of mind had deteriorated to that point.

Scott saw my despair one day as we prepared to sit down to lunch in Mitchell Hall, a cafeteria filled with four thousand noisy cadets. I hadn't said a word, but Scott looked over at me and said, so quietly I could barely hear him, "Chuck, I love ya."

In his letter to the Corinthians (I Corinthians 13:7), the apostle Paul tells us that love conquers all. Scott's simple declaration of love conquered my despair and gave me the strength to fight— to be worthy of a commission in the United States Air Force. Six weeks later, I threw my hat in the air in celebration of my liberation, thanks in large part to the love of Cadet Scott Wolfe.

I called Scott a few days after Katrina to check on his mother, Joyce, who still lives in Picayune. She had safely evacuated, and Scott was on his way to her house from Florida where he is stationed as a Special Operations C-130 pilot. He was loaded up with gas, roofing materials, and a chainsaw. I told him to be careful. I knew it was futile to attempt to discourage him from making the trip. He's as stubborn as I am.

A few days later, I called Scott again and told him that we were heading to DeLisle. He said DeLisle was only thirty-five miles from his mother's home, and that he would likely be there at the same time, since there had been too much damage to complete the repairs on his first trip. We made plans to meet.

Katrina had reunited four old friends and would soon breed new friendships and new bonds.

six

COMMUNITY

Eleanor, Ellen, and I left Memphis for DeLisle at 7:30 AM on Wednesday, September 14, seventeen days after Katrina struck. It reminded me of the last time I had driven a U-Haul truck, exactly like this one, with my yellow VW Bug in tow. It was March of 1992, and I was moving from Columbus, Mississippi, to Sacramento, California, because I had been selected to fly the U-2 reconnaissance aircraft. The U-2 has a crew of one and often flies over hostile territories. The motto of the U-2 is "Toward the Unknown," because you never know what possible danger or reward lies ahead.

This trip was different. I wasn't traveling alone in the U-Haul; I hadn't packed my worldly possessions, instead I had packed items for those who had lost all of theirs; I was driving toward Mississippi, not away from it; but one thing was the same—I was headed toward the unknown.

Eleanor drove most of the eight-hour journey to the Gulf Coast. Eleanor surreptitiously omitted Ellen from the list of authorized U-Haul drivers. Ellen loves to drive; she says that she has never been the same since getting her ticket to freedom on her sixteenth birthday. Unfortunately, the rest of us do not share Ellen's enthusiasm for her driving. It normally takes five hours, tops, to drive from Memphis to DeLisle, but we had three female bladders in the cab. We also had to stop in Jackson to pick up a few more supplies for the people on our relief list.

We began to see the effects of Katrina about sixty miles from the coast as we approached Hattiesburg, Mississippi. Large trees looked like weeds pulled from their roots and discarded by the side of the road. Roofs were torn up and tattered.

We were on the outskirts of nature's war zone. Katrina made landfall on Monday morning, August 29, 2005, packing sustained winds of up to 145 miles an hour. Those who were there are haunted by the sound of that wind. For many that sound is their most vivid memory of the hurricane.

"It was like a rushing freight train coming right into your living room."

"The sound of the wind was the worst, like the scream of an inconsolable insane woman howling right next to your ear."

People huddled with their loved ones (even if that loved one was a little dog) to stave off the fear and anchor themselves against the force of the wind. Katrina's winds whipped up a wall of water that washed away lives and livelihoods all along the Gulf Coast. Ranging from fifteen to thirty feet, it is the highest storm surge ever recorded in North America. The tidal wave violently flushed homes with a filthy water that collected more and more possessions, debris, and toxins as it traveled miles inland to neighborhoods that had never had more than a heavy rain.

Hurricane-force winds cut a path 120 miles wide, and wind damage extended a total of 250 miles. Ellen and I eventually drove the 120-mile span. Every home, every school, every business, every life had been uprooted to one degree or another. The declared federal disaster area covered 90,000 square miles, an area as large as Pennsylvania and Ohio combined. According to the Congressional Budget Office, Katrina destroyed an estimated 300,000 homes—over ten times the 27,500 homes lost in the four hurricanes in 2004 (Charley, Frances, Ivan, and Jeanne) or the 28,000 homes claimed by Hurricane Andrew in 1992.

One and a half million people were evacuated, and as of November 2005, more than 1,300 lives had been lost: 1,056 in Louisiana and 228 in Mississippi. One month after the storm, 363,000 people had filed for unemployment insurance in Louisiana and Mississippi. Twenty-five hundred people were still missing five months after the storm.

We stopped for gas near Hattiesburg for one last top-off, since we did not know what the fuel situation would be closer to the coast. I chatted with the man at the adjacent pump. He may have been fifty years old, but he looked about fifteen years older. He said that a few days before, the line for this gas station was two miles long and they would only let you have five gallons. He asked what we were doing with the U-Haul, and I told him that we were helping Katrina's survivors. His voice broke as he said, "Folks sure have been real good to us down here."

We pressed onward to DeLisle and met our hosts, Shantrell and Myrick Nicks, in an abandoned Chili's parking lot. They were waiting for us when we arrived. We all nervously smiled at each other, struggling to look like we were happy, but there was awkwardness to the meeting. I wanted to hug them, but I did not know them well enough to do that. Our differences were more apparent than our similarities.

Race was the most obvious difference. I had never felt the division as I did that day. The television images of predominantly poor black people desperately calling out for help had initially drawn out a sense of shame in me, and I still felt shame as I met Shantrell and Myrick. A voice in the back of my mind was saying to me, "So, you're going to come down here and deliver a few boxes of diapers to the poor, helpless black people and then go right back to your Upper West Side apartment in New York City. You're not going to do anything except make yourself feel better."

Shantrell and Myrick are not poor or helpless. They are professionals who achieved the American dream through education, not privilege. Myrick worked his way through college by serving in the U.S. Air Force Reserves. Shantrell is still paying off her law school student loans.

I respected Shantrell and Myrick even before I met them. It was Shantrell who graciously answered Ellen's peppering of questions on that flight out of Jackson. She then returned to the Gulf Coast earlier than she had planned to act as liaison between her community and our makeshift relief effort.

Shantrell surprised me with how quiet she was when we first met. I wasn't sure whether she was thinking the same thing I was—"Who do these people think they are?"—or if it was her nature. I later discovered that it was her nature. She has a quiet, gentle, non-anxious presence. She is steadfastly undaunted by life's challenges. Those challenges can range from meeting strangers to chasing after her eighteen-month-old son, from mediating a child custody battle to losing three-quarters of her home to mold.

Shantrell and Myrick's easy natures quickly silenced the skeptical voice in my head. Shantrell's smile began to speak for her, yet she stood with dignity and confidence as if she was about to represent her client before a jury of her peers. While she was wearing an oversize T-shirt and baggy pants, I sensed that this was not how she preferred to dress. Now I know that Shantrell normally does not go out in public (nor does she allow her children to go out in public) without being perfectly dressed. She and Ellen have this in common, along with an obsession for losing weight.

Myrick appeared to be the outgoing one in the family. He asked us about our trip and how we were doing, and if we had any trouble along the way. He was upbeat, as if we were meeting at a convention, not in a federally designated disaster area. I knew that Myrick was a vice principal. In my mind I became one of his students. I didn't want to disappoint him or get out of line.

Shantrell's cousin Belinda and her husband, Jesse, were there to meet us as well. Although they were in work clothes, they looked like they belonged on the cover of a magazine. Belinda is petite and speaks in an easygoing, gentle voice—except when sharing her private recipe for red beans and rice; her tone then becomes direct and serious. Belinda works with pre-kindergarten children. Jesse is now a high school football coach. He played safety in the NFL for eight years and still looked fit enough to walk back on the field. He towered over Belinda, but she appeared to call the plays in the family—softly.

Ellen had recruited two friends for our relief team, Jason Notini and Adam Sharon, who both met us in Mississippi. Ellen has known Jason since he was a child. He is the son of one of her close friends and was the only member of our team who actually possessed construction skills. He's a banker by day and master renovator by night and weekends. We would have been useless without him.

Adam Sharon has worked for Ellen full- or part-time for five years. We picked Adam up a few days later at the Gulfport Airport. I suppose I should say we rescued him from the air-port. When we arrived the local workers had told him about the mosquitoes in the area carrying deadly diseases, and that a few people had already been hospitalized. This explained why he had his long-sleeved shirt hermetically sealed around his wrists when we showed up in the middle of an afternoon with a humidity-adjusted heat index of 110 degrees.

Our crew, the three of us in our U-Haul and Jason in his rental car, followed Shantrell's crew to the Mt. Zion United

Methodist Church of DeLisle. We drove about twenty minutes west of Gulfport on Interstate 10 before turning off at Exit 24, Menge Avenue. I remembered that my friend Scott Wolfe had told me we would be taking Exit 24 to go to DeLisle. He said, "DeLisle. That's the exit I take to go to Mama's house—24, I think. I know that area well. It's a sweet little community. I hate it that they're sufferin'."

We made our way down a busy country road, then passed a micro-shopping center with two stores—a florist and a trucking company. Mississippi is flush with small and seemingly random entrepreneurial ventures. I'm sure the owners of these businesses conducted some sort of marketing research prior to breaking ground, but to a passerby, there doesn't seem to be any rhyme or reason for a florist to be located next to a trucking company with no other retail nearby.

These small businesses attract customers with small portable signs that litter the edges of Mississippi's roads and highways. The most popular version has a large neon arrow on top that points to the establishment and a message board that advertises the weekly special (12-Pak Bud Light $7.99) or any other information that the owner may want to announce, such as GO GATORS! There were no such signs on the road to DeLisle that day. I suspected that they were in someone's yard, or perhaps in the next county.

Myrick stuck his hand out the window to ensure that we knew the turn was coming up. I did not see any street signs. One traffic light dangled from its wires. We turned left on Lobouy Road. The sign was actually pushed over halfway down to the ground. Very few homes were visible from the road. It seemed as if we were in the middle of a forest that had been cut down by a giant lawn mower and left unraked. We weaved around the shredded branches.

Myrick slowed to a crawl and we made a hard left into a dirt parking lot. I thought to myself, *Are we at Mt. Zion already? Where's the town of DeLisle?* I was expecting a small church that sat in the middle of a little town surrounded by shops and homes. A scattered mess of donated clothes had been dumped in the front of the simple, honest, white cinder-block church.

I stopped the U-Haul as four young children, who were running around the church grounds, moved out of our way. I guessed that their ages ranged from five to twelve. They watched us pull into the lot with curiosity, but almost immediately went back to their task at hand, which was to put as many riders on a bicycle as possible. Two children were already on the banana seat. The oldest child was attempting to balance the five-year-old on the handlebars. I smiled at their determination. As I looked back, I saw that they had managed to make it work. The two riders were giggling and the driver/peddler focused on doing the work. The oldest child spotted them in case their experiment failed. I admired them, just as I had admired the children of Iraq who mysteriously retained the joy of play in the midst of war.

Observing this joy and resilience of spirit thriving in spite of so much loss has enabled me to understand why, when the disciples asked Jesus, "Who is the greatest in the kingdom of heaven?" he called a child to him and said, "Truly I tell you, unless you change and become like children, you will never enter the kingdom of heaven." (Matthew 18:1–3)

Rev. Rosemary and her husband, Rev. Theodore, came out of the church to greet us. I felt inadequate in the presence of Rev. Rosemary. Nothing in her manner should have elicited this response; she was warm and welcoming and smiled easily. Her hair was sprinkled with gray. Her glasses were thick, and the small, dark, rectangular frames made her look grandmotherly and professorial at the same time. But it was her posture that caused me to yield to the power of this diminutive woman.

She stood up straight as though she was intentionally pushing against the forces of gravity—not the physical forces of gravity perhaps, but the force of social gravity that Rev. Rosemary has undoubtedly resisted her whole life.

When Rev. Rosemary was assigned to the Mt. Zion United Methodist Church in 1996, there were eight people in her congregation. She was one of the first black women to be given a church along the Gulf Coast. She was still a full-time teacher, having taught for over thirty-three years. She tried doing both for a year—teaching school during the week and preaching on weekends—but the demands of being a pastor made it very difficult. When she announced her resignation to her school superintendent, he was sorry to see her go, but he said, "Well, I'm not going to get in a tug-of-war with God."

Rev. Rosemary is now pushing against the force of nature in the wake of Katrina's devastation. Her congregation has been wiped out. Every aspect of the material world that had existed for them before has changed forever. She now has a congregation of 136, and the church is growing daily as people come to her for comfort after the storm—not just spiritual comfort, but food, shelter, clothing, and love. The need is so much greater than her ability to provide alone, yet with all her might, Rev. Rosemary seems to be physically resisting the downward force of despair.

Her speech is deliberate, concise, direct, and perfectly enunciated. I hung on her every word, as few as they were—she listened more than she spoke.

There is one word, however, that she used more than any other—*community*. Rev. Rosemary paused to enunciate that word as though to elevate its importance above the others. Her voice deepened slightly halfway through the word, as if she was digging into her core not only to speak it, but also to command its existence into being. Her emphasis on this word above all others left me with no doubt that she believed survival was dependent on community. Her actions proved that she believed her mission was to build a refuge of community for God's suffering children.

Rev. Theodore, her husband, seemed to be the perfect partner for Rev. Rosemary. They were similar in size and stature and in their commitment to community. His hair was also sprinkled with gray and his wire-rimmed glasses gave him a collegiate look. Like Rev. Rosemary, Rev. Theodore had been a teacher. He taught music for twenty-one years before becoming an ordained minister in 2003.

He had his own beautiful church, St. Paul United Methodist, before the storm. It was located a few miles away in Pass Christian, Mississippi. "The Pass," as they call it, is another rural community. The town, along with St. Paul Church, was destroyed. Now he and his congregation are worshiping at his wife's church.

Rev. Theodore was much quicker to smile than his wife, and he appeared to carry Katrina's burden a little less heavily in his heart. He laughed and joked about his daughters and grandson and frequently spoke of God's blessings for the past, present, and future.

Rev. Theodore also saw himself as God's servant, but it seemed that he believed God was going to do the work with or without his help. Rev. Theodore would do his best and God would do the rest. He spoke often of the Lord and how the Lord blesses His children. He embodied the words of a song I used to sing as a child: "The Joy of the Lord."

Rev. Rosemary's first words to us were, "You all have blessed us so much by coming. Let us help you unload."

I was at a loss for words. Our truckload of supplies was so insignificant compared to what her community was suffering. In spite of what my head was telling me, my heart had heard Rev. Rosemary's words. *You have blessed us so much by coming.* The blessing was not only in what we brought, but also in our presence there with them.

I focused on helping unload the truck. Manual labor was a welcome diversion, enabling me to avoid the emotion that surfaces at a time of loss, even the losses of complete strangers.

The Mt. Zion Church structure itself came through the storm with little damage. Everything appeared to be in working order: the modest tile roof, the white cinder-block walls, the plain windows, even the trees that still provided shade and relief from the heat. Other than piles of donated clothing outside the entryway and the makeshift barbecue grill being used to cook for the community, no one would have known that anything was out of the ordinary at Mt. Zion.

Inside, however, the church looked like a Red Cross relief center. Cases of water were stacked neatly along the walls and in the only hallway in the church. The small church kitchen was acting as a pantry, with dried and canned goods stacked everywhere. The back pews were reserved for cleaning supplies and toiletries. The large cross, the small pulpit, and the pews were the only reminders that I was in a church.

Rev. Rosemary was playing grandmother to the children of the neighborhood as she directed our efforts. I counted ten children. She was looking for a basketball for one little boy who had been asking her for one. He looked about three years old and had beautiful blue eyes. The children ran in and out of the church and tugged on Rev. Rosemary for attention. She graciously met their needs as she spoke with us. She reminded them to be sure to close the door to keep out the bugs and heat. "Now, children, we need to keep that door closed if it's going to remain cool in here." They understood. It felt like a steam room outside.

Rev. Rosemary called these children her "Children of the Storm." They had not spent much time with her until Katrina swept away the basics of survival—food, water, shelter—along with the basics of childhood, fun and games. She was determined to fill all their needs as she welcomed her Children of the Storm home—home to Mt. Zion.

While the church came through the storm with "little damage," as they say in the Gulf, its congregation did not fare as well. After we finished unloading half the contents of the truck, the whole crew climbed into the church van to tour the devastated area that included DeLisle and parts of Pass Christian, an area with a population of fewer than ten thousand people.

DeLisle is a community in the truest sense of the word. Many of its residents live on land that has been in their families for generations. The median income in Mississippi is approximately $23,000, and yet people somehow manage to buy homes and raise families as long as they have steady work. But most members of this community have no margin for error. They can't replace a car battery, let alone recover from a category four hurricane. They do not have the magic wand that I have— credit. While the phrase *the working poor* is accurate, it fails to fully describe what it means to be a poor working American citizen. *Working poor* doesn't afford them the dignity they have earned, and it doesn't fully express the courage, pride, and hope they bring to their lives and communities.

Katrina's thirty-foot tidal wave rolled through a neighbor-hood within a quarter mile of Mt. Zion that was supposedly on high ground, thought to be so secure that the residents of Pass Christian had driven their cars there to save them from water damage. Katrina thought differently. As her tidal surge pushed through, people ran from house to house, trying to get to higher ground. Eventually, sixty-five people ended up on the second floor of an evacuated home. Their neighbor, who had left town, was grateful that they were able to get into her home and be safe.

As bad as DeLisle was, Pass Christian was much worse. Some of the buildings remained standing, but they had shifted from their foundations. Three homes next to Rev. Theodore's church had been blown from one side of the church to the other. The residents of the Pass did not have full access to their property when they tried to return home because of the environmental biohazards and the ongoing search efforts. Most of them were living with relatives and friends in the surrounding area, many of whom had their own flood and wind damage.

DeLisle's residents had access to their homes, but the water damage and heat and humidity had caused a proliferation of mold that made their homes uninhabitable. Some stayed in their homes anyway and endured the mold; others pitched tents outside or slept in their vehicles.

It was the hottest part of the day, but I noticed that everyone was sitting or working outside. Some were tinkering with their cars or moving debris from one pile to the other. I realized that most had no reprieve from the heat. They were living in a steam room 24/7. They had no air-conditioning, no washers, no dryers, no fresh water, no gas, no refrigeration, and no fresh food. They were "camping," and the only camping supplies they had were ones that other people delivered to them. Underwear and bug spray were at the top of nearly every individual supply request list that Shantrell and Rev. Rosemary collected.

DeLisle and Pass Christian are racially mixed. Katrina was colorblind. Cars had been swept by the water and carried on top of other debris; trees had crashed through roofs; FEMA and insurance identification numbers were spray-painted on plyboard, the only way to identify a home's address. As we drove by I saw a woman standing on her wraparound porch, now half missing, with a broom in one hand and a cell phone in the other. She was sobbing. I had seen the brunt of Katrina's destructive force on the material world, but this was the first time I saw her fury break a human heart.

seven

ONE DAY AT A TIME

Rev. Theodore and Rev. Rosemary drove us around in the church van for about forty-five minutes. I could not imagine how these people could get up day after day to face a future that was almost entirely out of their control. The simple antidote for dread and anxiety that I have relied on since I was a child rushed to mind: "One day at a time." Perhaps, in some small way, we had helped some of them get through just this one day.

It was after 5 PM and we had one more stop to make. Shantrell had arranged for us to distribute supplies in Gulfport, a much larger town than DeLisle or Pass Christian. With a population of seventy-two thousand people, Gulfport, like many towns in Mississippi, is approximately 60 percent white and 40 percent black. Shantrell and Myrick live in Gulfport, and Shantrell was raised there. It seemed as if she knew everyone, black and white. In an attempt to accomplish the most good for the greatest number of people at one time, she concentrated on one particularly hard-hit neighborhood named Forest Heights.

Forest Heights, nicknamed "Turnkey," was not in a floodplain or near the beach, yet was drenched when nearby waters overflowed. Most of the homes were in more than four feet of standing water. Several residents had to be rescued by ad hoc first responders in small, privately owned fishing boats. Forest Heights consists of about two hundred modest three-bedroom homes. The racial make-up is predominantly African American.

The development, built in the 1970s, was one of the first affordable housing programs in Mississippi thanks to the combined resources of the Department of Housing and Urban Development, the Ford Foundation, and the homeowners themselves. Its name was inspired by Dorothy Height, the tireless public servant who drove one of the first stakes into the heart of racism in 1928 as a high school student in Rankin, Pennsylvania. Dorothy's new principal had forbidden her to lead the singing at the school assembly. Her white classmates refused to sing until the principal motioned for Dorothy to take the stage and a chorus erupted. Shantrell later told me that the homeowners in the neighborhood had recently celebrated paying off their thirty-year notes.

Our small convoy—those of us in the U-Haul, Jason, and Shantrell's group—arrived at Forest Heights early, so Shantrell led us on a tour of the neighborhood. The residents were outside, attempting to clean and dry out the contents of their modest ranch-style homes. The air was so humid you could practically drink it, so I assumed their efforts were in vain. The combination of wind and water damage made the whole neighborhood look like a giant, filthy gutter. I later learned that the interiors were no better. Mold had already crawled up the ceilings like ivy.

Shantrell told people to meet us at 6 PM in the parking lot of a closed community center adjacent to Forest Heights. As we pulled into the parking lot, people immediately began to show up, as if the U-Haul was a giant magnet. They asked anxiously if they had to have their name on a list to get these supplies. I sensed their desperation for the first time. I said no. We had plenty in addition to the bags they could see with names on them. Their eyes lit up, and they politely asked for the specific things they needed.

The neighborhood children jumped right in to help. Well, except for our first customer, who ran up to the truck and asked, "You got any TVs?"

"No," we told him.

"Well, then can I have a Gatorade?" he asked, and left.

Few people had television service for almost a month after the storm, but I suppose this young man did not know that, because he had no TV or any other household appliances. Radio was the only media available, and it was an invaluable source of information, from public safety announcements to job opportunities.

Many of those who came were still wearing their work clothes. Their jobs had barely enabled them to make ends meet before the storm—I worried about what would happen to them now.

Eleanor and I loaded the last package of diapers into a car belonging to a woman with a friend and four children. The children were wearing beautiful smiles even though their clothes looked as if they had not been changed since the storm hit. They were piled into the backseat of the car, nudging each other, giggling, and holding the little toys that Eleanor had bought in Memphis as though they were gold.

Their car was a mess, with rust spots all over it and several dents. The velour interior lining was hanging down from the roof. The mothers of these small children tried to hide their anxiety with gratitude as they got into the car, smiled, and said, "Thank you." One of them started to ask a question, then hesitated. Suddenly she asked softly, "Is there any way I can get my name on one of those lists for the next time?"

I asked her what she needed and she said, "Baby clothes." The other woman said, "A mop." Eleanor reached into their car and gave the women some money that Miles, Eleanor's teenage son, had given her to give to someone. Shantrell came over and took their names, and the next day she and Ellen collected supplies for those women and several others who had shown up the night before with similar requests.

We said our quick good-byes to Shantrell, Myrick, Jesse, and Belinda. We were scheduled to meet Eleanor's cousin Frankie that evening at 7:30. Eleanor had arranged for the three of us to stay with Frankie and his wife, Carol Ann, who lived in Ocean Springs, Mississippi, about twenty-five miles east of Gulfport. Ocean Springs is a town of approximately eighteeen thousand people. Unlike DeLisle, Pass Christian, or Gulfport, almost 90 percent of Ocean Springs is white, but they also felt Katrina's wrath—that is, most did. Frankie's home had managed to escape flooding or significant damage. He boasted that his home was on the highest plain of ground in the entire neighborhood. A large oak tree was down in the backyard, but otherwise his home was sound.

Frankie met us at a parking lot someplace off Interstate 10. By this point, my mind was starting to filter out as much content as it could. I had already seen more than my brain could safely process. Like an airplane electrical system, it was starting to "load shed," meaning that nonessential components were dropping off-line to ensure that my critical functions, such as breathing, could continue.

Frankie led us through another Katrina-ravaged community. Ocean Springs had been a beautiful little town—the kind of place where you could go on a weekend and have a nice lunch at an outdoor café, then mill around the antique shops and art galleries. Large trees that once shaded the streets were now on the roofs of the wrecked cafés and shops.

We pulled into a large parking lot to leave the U-Haul for the night. As we drove through Frankie's neighborhood, I was glad that we hadn't tried to bring in the U-Haul. It was already dark, but I could see that power lines were still dangling and piles of debris littered the streets.

We filed into the house—Jason, Ellen, Frankie, Carol Ann, Eleanor, and me. I was greeted with friendly slobber from a big dog that looked like a mix between a Labrador retriever and an I'm-not-sure-what.

"That's Ulrick," Frankie said. "He won't hurt ya. Ol' Ulrick rode out the storm with me. Carol Ann left town."

I thought to myself, *Only a dog would be so loyal to a crazy ol' fool.*

There was at least one cat (that I knew of), but I tried to avoid it even more than I tried to avoid Ulrick. The cat figured out that I was not a friend of the feline species. (Cats seem to have a great deal of emotional intelligence when it comes to understanding which people don't like them.) Ulrick, on the other hand, continued to shower me with slobber. I love animals, but not their hair, and cat hair is the worst. I later discovered that Jason is also highly allergic to felines. He said he about panicked when he saw the cat food dish on the front steps as we walked in. To make matters worse, the cat's favorite hangout was the couch that would be Jason's bed.

Jason never said a word about his allergies while we were at Frankie's house. He and I had seen too much devastation to worry about our own temporary problems that night.

Ellen put aside some of her allergies as well—political ones, that is. Frankie was prepared to engage in a friendly game of Red State–Blue State and asked Ellen if she wanted to see his George Bush screen saver. Ellen was unwilling to engage. That's when I realized how tired she was.

Frankie held court over the dinner Eleanor had brought from

Memphis. He told us how he and Ulrick rode out the storm together and showed us the map he had used to plot Katrina's course, complete with waypoints and military times marked along the path. The plotting stopped at a few hours before Katrina struck the Gulf Coast. "What happened after that?" Jason asked.

"Ah, I stopped plotting when I figured out she was headin' right for me." There was a hint of "never gonna do that again" in his voice as he said it.

Frankie was proud that he had bounced back, as resilient as ever, after the storm. He was determined to return to the life he loved, and there are few greater loves in the South than college football. Frankie is an alumnus of Mississippi State University in Starkville (pronounced "Starkvull"), home of the Mississippi Bulldogs (the Dawgs). The Dawgmobile, as he calls his white Ford Taurus (wallpapered with Mississippi State, George Bush, and Haley Barbour stickers) had a flat tire from the storm, but that didn't stop Frankie. After the hurricane, he slapped on the temporary miniature tire and headed north for the four-hour drive. "Had to see my Dawgs play," he said.

Ellen decided this was a perfect time for a break in the action and announced that it was time for bed, and we said our good nights.

The next morning I awoke to the smell of muffins baking. Frankie had made coffee and breakfast for his guests from the North, and he was already touring Ellen around the neighborhood so she could do her morning drive radio shows.

I went for a run along the bay, where I came across a marina that looked like a bathtub after a three-year-old has flung toy boats everywhere. The beautiful antebellum homes had been crushed. High-back leather chairs and big-screen TVs were out in the street to be picked up. Families were wandering around their properties seemingly in shock, searching for some piece of their past among the tons of debris.

Afterward, we packed up our things, took a few group photos and said good-bye to Carol Ann, Frankie, Ulrick, and Eleanor, who had to fly back to Memphis that afternoon.

Jason, Ellen, and I drove back to DeLisle to meet Shantrell and Rev. Rosemary. As we planned our day, several adults came in and out of the church. Some attended Mt. Zion or St. Paul, but most did not. They were from the neighborhood and needed help—teachers, retirees, casino workers, janitors, home care-givers—all normally self-sufficient, working Americans.

A retired couple came into the church and said that their brother had gotten into a terrible car accident while trying to drive nonstop from California to pick them up. "He's okay now," they said. We heard similar stories from others whose relatives raced to help them. This same couple joked that their son had bought them a cell phone, but it was a week before they figured out how to use it, so no one knew if they were okay.

One of the women who lived next door to Mt. Zion came in to see if anyone could help her make a phone call to her bank. FEMA was willing to grant her $2,000 in emergency relief, but they needed a bank account number and she did not have anything left after the storm with an account number on it. Jason handed her his cell phone. She couldn't get through so Jason started trying.

The woman was clearly pregnant. "When's your baby due?" I asked her. The question transported her out of her state of anxiety and a big smile spread across her face.

"November," she replied. I congratulated her and she smiled again, only this time with a subtle note of pride unique to expectant mothers—as if they instinctively sense the power of the gift they are carrying.

Jason managed to get a dial tone and handed the phone to her. She got through but then hung up, looking dejected. She smiled in gratitude as she handed Jason's cell phone back to him.

"They won't give out that information over the phone," she said in little more than a whisper.

"Where's your bank?" I asked. "We can take you there."

Again she smiled gratefully. "Keesler Credit Union," she said with regret.

Keesler Credit Union is on Keesler Air Force Base. Katrina closed Keesler. This expectant mother had $2,000 waiting for her that she desperately needed, but no way to get to it without a bank statement and without a bank.

Some people came into the church to tell of help on the way. Cell phone coverage was still sporadic, so people went from house to house and church to church to spread the word.

"They're giving out FEMA trailers on Menge Avenue," one woman came in and announced. "You just need a number. Call everyone you know. They aren't going to last!" She explained to people that they would need to have their FEMA number with them.

When you first apply to FEMA, you are given a number. This number is the currency for disaster relief. You cannot do anything without it. Unfortunately, getting a FEMA number is not easy for people who have no phone or internet connection. The woman told us it was so hard to get through to FEMA that it took her eight days to get a number.

Some people drove to FEMA centers in neighboring states. "People are goin' all over," the woman said, "some to Alabama, some to Georgia, even all the way to Florida to get to FEMA. I wish they would hire us to work for FEMA. I can do the work and you don't need to find me a place to stay. I've done that for myself." This woman was a casino worker. Later in the morning, we saw her working behind the counter at the local gas station.

Before we began buying more supplies, we decided that we should try to find a place to stay for the night. I called to see if my friend Scott Wolfe was back at his mother's in Picayune. I knew his mother would make room for as many people as we could bring to her home. I tried all day to contact him, but the cell phones were sporadic and the regular phone lines were still down.

Shantrell insisted that we stay with her, but we knew that her home was damaged, and we didn't want to put an added burden on her. We made a "plan Z," a plan of last resort, to sleep in the back of the U-Haul.

As we went from motel to motel and realized that they were either closed or booked for the next five months, we started to rethink Shantrell's offer. Even the swimming pool at the Holiday Inn was occupied: the dolphins from the local aquarium had been rescued by the Navy, and they were staying in the hotel until they could find a permanent residence.

Shantrell persisted. "My kids are with my in-laws in Hattiesburg," she offered. We thought about the heat and the mosquitoes and the fact that the only bedding we had was Jason's pillow. We moved into the Nicks's home that same night. Adam arrived the following day so he ended up with the "Ladybug Room" (three-year-old Taylor's bedroom). We stayed the rest of the week. The Nicks's Bed and Breakfast has since become a second home to us.

Staying with Shantrell and Myrick was a highlight of that first trip. The six of us sat around, talked about the day, and laughed as if we had known each other all our lives. Laughter is the best antidote for heartbreak. Sarcasm is also very helpful, but it isn't as effective or as much fun unless it leads to laughter.

One night Jason and I got to talking about how special the people were whom we had met. "Can you imagine what they could accomplish if they had just *one* person who invested in them and mentored them—if they had one tenth of the opportunity that I have had?" I asked.

In retrospect, my rhetorical question was unfair to those survivors. They *had* achieved great things. They had homes, children, and grandchildren. Most important, they had built a community they loved. Katrina had blown away their material world, but not their love for community, and not their love for each other.

eight

COURAGE UNDER FIRE

My mother was a single mother for half of my first seven years. Those years were probably some of the most difficult of her life, but for my brother and me, they were some of the best.

While I am eternally thankful for my stepfather, James Johnson (he and my mother have been married for over thirty-three years), I also have very fond memories of life when it was just my mother, my brother, and me.

My mother had gone through two divorces before I was seven, but despite her own pain, she made having fun a priority. When we were living in Albuquerque, New Mexico, she would pick us up from daycare, still in her dental hygienist uniform, and take us straight to the Big Slide.

I was so proud of my mother. She was a professional. She was beautiful. She wore a uniform that made her seem special compared to the other moms. I guess that's when I first started to like uniforms. She must have been exhausted from being on her feet all day, but she would march us up and down and up and down the multiple steps to the top of the slide with our burlap potato sacks in tow—all so we could make our stomachs turn upside down and giggle until we could hardly walk. We aimed to go fast enough to be airborne as we topped the bumps. Perhaps that's where I picked up my love for speed.

Once a week she would take us to our other favorite place, Bob's Big Boy, so we could have the best food ever invented— hamburgers and french fries. She had very little money, but she insisted that we have every educational opportunity possible.

In those days, that meant you needed the Encyclopedia Britannica. I was too young to take advantage of the gift, but just having those shiny new books loaded with interesting pictures made me feel special.

Until I went to Mississippi, I hadn't spent much time with single mothers other than my own. I'd forgotten how committed they are to their children and how determined they are to provide the love, in some cases, of two people.

There are many single mothers in Mississippi. I'm sure there are several explanations for this, none of which really matter to a mother who is trying to raise her children as best she can—even under the impossible circumstances created by a hurricane that destroyed what little livelihood she had.

Sheila lived across the street from the community center where we unloaded the relief supplies. She had a small home of approximately six hundred square feet; in her front yard a large tree had been uprooted. We asked her if we could go inside to see the water damage.

Sheila was so proud to show us her home. She told us that she rode out the storm in that little house, though she agreed that she would never do that again. She had stripped out the carpeting and was trying to dry out the mold with fans, but she was losing the battle. Mold was already crawling up the walls. It was so bad that our lungs felt heavy after only five minutes inside with her.

Jasmine, Sheila's seven-year-old daughter, was our biggest helper when unloading the truck and making the supplies accessible to everyone. She followed us into her home. I noticed that Sheila had bought Jasmine the modern equivalent of a set of encyclopedias—Jasmine's computer had survived Katrina and she had a school software program running on it.

Sheila and Jasmine are now staying with Sheila's mother in

her small house. It will take at least $26,000 to repair Sheila's home. In addition to the mold, the entire foundation has shifted. Sheila works but cannot qualify for a private loan under the current circumstances. She was insured for hurricane damage but her insurance company said that the damage did not exceed her deductible. Her deductible was $300. She plans to appeal the decision, but in the meantime she's stuck with waiting to see what FEMA and a Small Business Administration loan will provide.

The next day we went to the Lockett Williams Mortuary, which was acting as a community supply depot. It was centrally located and owned by Rev. Rosemary and Rev. Theodore. Their daughter, Sonya Williams, another single mom, manages the mortuary. Her father calls her "my right hand, left hand, and everything else."

Sonya's five-year-old adopted son, Shedrick, was born to a twelve-year-old girl whom Sonya had mentored. The middle school principal hoped that Sonya could make a difference in this troubled girl's life. She was a straight "A" student, but the circumstances in which she was growing up were hardly conducive to developing into a responsible adult.

I once attended a seminar on mentoring sponsored by United Airlines. The man leading the course said there are five things a child needs to become a fully functional, contributing member of our society:

1. An adult in his or her life who is irrationally committed to his or her welfare. (He did not say an irrational adult who should be committed!) This adult does not need to be the child's parent. It can be a grandparent, aunt, uncle, friend of the family, or mentor.

2. The basic needs for food, shelter, and safety must be met.
3. A child needs to learn to read, write, and do arithmetic.
4. A child needs to learn skills that can later be applied in the workplace.
5. A child needs to be able to give back to his or her community.

Sonya was the adult—the only adult—irrationally committed to this young girl's welfare. Sonya took responsibility for the first step on this child's road to becoming a fully functioning adult. Despite Sonya's commitment, that child had her own child at the age of twelve.

Sonya continued to mentor the young mother and helped support her and her new son. Ultimately, the young girl was not ready for motherhood. When a judge asked Shedrick's birth mother what she would like to have happen to the child, she said she wanted Ms. Sonya to raise him. The judge called Sonya and she picked up Shedrick from the courthouse when he was four weeks old. Thanks to Sonya's selflessness, Shedrick will get past the first step on the road to full development. He has not just one, but a whole community of adults irrationally committed to his welfare.

One week while I was in Mississippi, Shedrick earned the Student of the Week award at his school. The award has an additional privilege aside from the prestige that comes with such a distinction: each Student of the Week gets to bring a treat for his or her classmates. Most children bring cupcakes, Fruit Roll-ups, or a similar treat. Not Shedrick. Although he is only five years old, he is sharp enough to realize that his class had not had much in the way of real food to eat since Katrina, and this required big thinking, even among Gulfport's smallest citizens. When Sonya asked him what treat he wanted to take to school he said, "Collard greens, fried chicken, and macaroni and cheese." Sonya was a little taken aback, as the request was made at the breakfast table on the day the treat was to be produced by the Student of the Week.

Collard greens, fried chicken, and macaroni and cheese are not on the "Katrina Menu" that the few existing restaurants serve. Sonya works and everyone she knows works, including her mother and father, so that eliminated the possibility of recruiting others.

Sonya, like mothers everywhere, found a way to do the impossible. I asked Shedrick what made him want to feed his class this big meal. At first he said, "I don't know." But after he thought for a moment he said, "Well, my teacher loves fried chicken." Perhaps the five-year-old Shedrick instinctively knew that his teacher and class *needed* a real meal. He is already accomplishing the fifth step on the list to becoming a "fully functional, contributing member of society." He is determined to give back to his community with the help of the person who has always given to him.

The first time I met Sonya she looked visibly stressed. She had been arranging for the burial of some of Katrina's victims whose families had no money to bury their dead. She called the Batesville Casket Company to ask if they would consider donating a few caskets. They refused, saying they gave to the Red Cross. Sonya did not tell us this, but we understand that her family ultimately paid the burial costs so these families could give their loved ones a proper burial. While FEMA lists burial expenses on its website as one of the potential disaster benefits, it is up to the family to pay the expenses and then request reimbursement. The last thing

Sonya intended to do was collect money from these destitute people.

Four women lived next door to Mt. Zion Church in two shacks and a tent. One of the mothers looked like she might be in her forties, but was probably not yet thirty. She had the same beautiful blue eyes as Rev. Rosemary's little child of the storm who was trying to find a basketball. The four women had two toddlers, and it appeared that the women were sharing the property with two older men, probably related to them. They were all living in these two shacks and the tent, which sported a Rotary Club emblem. I assumed it was a donated temporary shelter. Jason, Myrick, Anthony, Adam, Rev. Rosemary, Shantrell, Ellen, and I all went over to see if we could help them out.

The shacks were only partially standing. There was burned trash everywhere and the smell of urine was overpowering. One of the toddlers was running around with no shoes on—she didn't have any. One of the two older men was still wearing his golf course groundskeeper uniform, even though he had not been to work since the hurricane had closed the course two weeks earlier. He told us that he liked working there because he could walk to work; they didn't own one car between them.

Shantrell, Ellen, and Rev. Rosemary sat with the women and talked while Jason, Myrick, Anthony, Adam, and I nailed tarps on the roofs. I gained a new appreciation for the expression "hot tin roof." Jason made sure we didn't fall through and directed our efforts. I was shaking. I'm not afraid of heights, but I discovered that I'm afraid of falling through a tin roof. We had been up since 3:00 AM and had driven to New Orleans and back, so I was not feeling "Fully Mission Capable," as we used to say when determining the maintenance status of a U-2 airplane.

The two older men, probably in their late forties or early fifties, were drinking beer. They were very thin. The man in the golf course uniform reminisced about Hurricane Camille as we took a break from the roof work. "I was here for Camille. I'll take Camille any day. Camille wasn't hardly nothin'. Katrina, well, she's nothin' but a bitch."

Rev. Rosemary shot him a look and scolded him for using foul language, as if he were a schoolboy.

"I know, but what am I supposed to say? I can't take it anymore. I just want to leave here forever."

Various young men came and went in cars, and one young white man showed up while we were putting one of the tarps on the roof. It looked as if he was conducting some type of buying or selling transaction with the other young men, who then left quickly. Once I realized what the young men coming and going were doing, I was ready to write off the whole compound. I wanted to cry for the children who would have to grow up in this environment; they didn't stand a chance. As I hammered nails through the tin roof into the few solid beams left, I thought, "This whole situation is hopeless—the roof, the homes, the women, the children, the old men—it's a rat hole."

As I got down from the roof, however, my point of view changed. I noticed that the mothers were playing a new game of Scrabble they had borrowed from Mt. Zion. They were better at the game than I ever was. They didn't even try to make up words like we used to do. I was humbled. Once again my narrow, ignorant stereotypes had been shattered. The next morning those mothers were sitting in Mt. Zion's Sunday church services with their children.

nine

AMAZING GRACE

Sunday morning, September 18, 2005, was supposed to have been the 113th birthday celebration of the Mt. Zion United Methodist Church. Katrina postponed the celebration indefinitely, but Katrina could not stop Sunday church services at Mt. Zion. In fact, Mt. Zion's sanctuary was made available to two other churches so they could hold their own services as well.

Mt. Zion was packed with new faces, and there was standing room only. The usual congregation was there plus those from St. Paul United Methodist Church, along with people from the community who had been fed and clothed by the church. Now they sought another kind of nourishment.

Based on my previous church experience, I had warned Ellen and Adam, who are both Jewish, and Jason, who is Catholic, that this service could last for more than two hours. Ellen had to verify this with Rev. Rosemary the day before services by asking, "Cholene said that your church service lasts over two hours. Is that true?"

I wanted to crawl under the table. I should have learned by now that anything I tell Ellen might be broadcast to a wider audience. The church ladies started laughing.

"Oh, not that it's a problem for me, I mean, uh, two hours is fine with me. I just wanted Ellen to be prepared."

"Oh, no, an hour and fifteen minutes at the most," Rev. Rosemary reassured Ellen.

"You must be thinking about the Baptists," one of the church ladies said. "You didn't hear that from us," another lady added, and she started giggling.

I wasn't sure what to expect; I had lost my ability to connect with a church service in any meaningful way. Church had become a ritual or a way to pay some type of respect to God, not an opportunity to replenish my spirit. The last time I had felt the presence of God in a religious service was during Yom Kippur, the Jewish Day of Atonement. It was during the Neilah service, which concludes the Yom Kippur observance at sunset. It is one long collective prayer, both spoken and sung.

The power of this prayer is overwhelming; all Jewish people have one final opportunity, before the end of the Jewish year, to ask for atonement in the same language and in the same words that they have used for over three millennia. I imagined all the Jews over the course of history who have offered this prayer, some under circumstances that I cannot even fathom; yet they remained faithful to God and faithful to each other.

The church of Mt. Zion had already broken my stereotypes about faith-based organizations—it had shown me a different way. Mt. Zion was not simply in the community; it was at the center of the community. Over the course of the week, these people had expressed their love for God through their love for one another.

Some of the people working on behalf of Mt. Zion did not even attend Mt. Zion. Shantrell and Myrick lived in Gulfport and were not members. Most of Mt. Zion's congregation is middle-aged, which is typical of the local churches. I asked Shantrell one day why it was that younger generations were not as avid churchgoers as their parents had been.

"I don't know exactly. We don't go to church that often, but we are going to get more involved now. We had wanted to for a long time. For my parents, church was *all* they had. It was their social life. It was their spiritual life. It was everything. Now there are just so many other opportunities for people. It's not at the center of their lives anymore."

On this Sunday, Mt. Zion had a special visitor, Bishop Hope Morgan Ward of the Mississippi Area United Methodist Episcopacy. The Bishop was based in Jackson, two and a half hours north, but had been doing relief work near the coast.

The readings included the same scripture that had given me hope for Katrina's survivors two weeks earlier: "So the last will be first, and the first will be last." (Matthew 20:16) I felt affirmed that I was where I needed to be, and yet I knew I had a long way to travel with my new friends. I had not yet seen evidence of the last being first.

The Old Testament reading was Exodus 16:2–5. The whole congregation of the Israelites complained against Moses and Aaron in the wilderness. The Israelites said to them, "If only we had died by the hand of the Lord in the land of Egypt, when we sat by the fleshpots and ate our fill of bread; for you have brought us out into this wilderness to kill this whole assembly with hunger."

I could feel a wave of empathy throughout the congregation as the Old Testament came to life thousands of years later in the land of Mississippi.

The bishop stood in the middle of the platform with no notes, and explained that it had been difficult to choose the scripture this morning, but that she could not resist the Old Testament story. The congregation chuckled softly.

The Israelites were unhappy, she explained. They were not used to eating manna. She said there had been speculation about what, exactly, manna was. "Like MREs," one person spoke up. The bishop laughed at the MRE comment and pantomimed opening an MRE. "What *is* it?" She then said, "God provides, even though what is provided may not be what we expected or may have wanted."

She told about a local church that was feeding approximately five thousand people a day, but was running out of food. "We only have about ten pounds of chicken left and these people are going to come again today and expect to be fed," one of the people said.

"First we're gonna cook the chicken. Then we're gonna pray," someone else said. And so they did. Around midmorning, a truck unexpectedly showed up with two thousand pounds of chicken. But then they didn't have enough cooks.

A visiting men's Bible study group from Georgia who had brought their chainsaws to clear trees for people was nearby. "We need you to cook chicken," someone said. The men explained that they had come to cut trees and they were all set to do that and they weren't cooks. "You can cut trees and help out a few families, or you can cook chicken and feed thousands." These men whipped out the waterless hand soap and got to cookin'.

Sometimes God has a plan that is different from our plan, the bishop explained. But ultimately, God delivers.

"The first twenty days after a disaster is called the emergency phase. We are just now at the end of the emergency phase. Ten times that equals two hundred days, and that is the relief phase. Ten times that equals . . ." As she paused, one of Rev. Rosemary's Children of the Storm yelled out, "Two thousand!"

"That's right—two thousand days of recovery. We have a long way to go, but God is with us."

At the end of the service, Rev. Rosemary announced that she had a special treat for the congregation. "Children, children, come on up here." She said, "I call these my Children of the Storm. They have worked very hard for you all today."

About a dozen children from ages two to twelve took to the stage and with big smiles on their faces sang, "This little light of mine, I'm gonna let it shine, let it shine, let it shine, let it shine."

I could not hold back the flood of tears as they sang. I had seen their homes that were not fit for animals, let alone children; I had seen them running around in the dirt wearing no shoes and filthy clothes; I had seen their mothers come to the church to get food to feed them. Yet now they were on the stage with their hair perfectly combed and wearing clothes that had belonged to someone else's children, singing their hearts out with smiles on their faces that could light up a city.

Three weeks later I attended Sunday services at Mt. Zion again. We were twenty-one days into the two hundred days of the "relief phase" the Bishop had spoken about. But something was missing. There was no sign of "relief." It had been six weeks since the storm and only a few people had seen an insurance adjuster or a FEMA inspector.

As the winter rapidly approached, few had seen or received the relief that would ultimately enable this community to recover. The stress was beginning to wear on the nerves of the strongest members of Mt. Zion.

Six weeks after the storm, Shantrell called. "Do you remember the musical director at Aunt Rose's church? Ms. Bowser is her name." Of course I remembered her. Ms. Bowser was the retired fifth-grade teacher who sang "Amazing Grace" with such passion that she made me cry.

"Well, I think Ms. Bowser is on the verge of a nervous breakdown."

"What happened?"

"She just showed up at my office yesterday. Her hair was

sticking out every which way. She looked like she hadn't been sleeping. She was a mess. She wanted me to write a letter to her insurance company."

Ms. Bowser, her husband, and her brother-in-law were all living with her daughter, Sonya, and her family. The Bowsers' home in Pass Christian had been destroyed. Ms. Bowser had flood and hurricane insurance, but her insurance company ignored her calls and did not say they would send anyone out to assess the damages. Many people had been told their damages were not covered by insurance. Ms. Bowser could not imagine a scenario where neither flood nor hurricane insurance would cover her damages. She may have been on the verge of a breakdown, but she had the presence of mind to know that she needed a lawyer.

Shantrell helped her write the letter on plain paper instead of her legal letterhead. Shantrell did not want to start a legal fight with the insurance company—she is by nature a peacemaker. She never raises her voice or allows herself to get drawn into the emotion of a fight. Sometimes I wonder how she ended up in the legal profession.

A few days later Ms. Bowser again stopped by Shantrell's office unannounced. Shantrell said she looked worse this time than she did on that first visit. When Shantrell told Ms. Bowser that she had been trying to get in touch with her, she said, "I'm so sorry, Shantrell. I think I'm just losing my mind. I forget things all the time."

Ms. Bowser applied for a FEMA trailer for temporary shelter until she could find a new permanent home, assuming that the insurance company eventually would pay her claim. She was one of the first people after the storm to apply. Unfortunately, the person who took her request did not enter her name into the system. As a result, she had to stand in line again to reapply and get the coveted FEMA trailer number. Her name went to the bottom of the list.

I saw Ms. Bowser in church the Sunday after her visits to Shantrell's office. She played the piano and led the music with the same level of passion as the first time I saw her. I thought about all she had been through and listened to her sing "To God Be the Glory." She sang with such conviction there was not a dry eye within earshot of the Mt. Zion United Methodist Church.

Ms. Bowser's solo was followed by a flute solo of "Amazing Grace" by Rev. Theodore. Shantrell calls Rev. Theodore "Uncle Skin." I asked her where that name came from. "I don't know. We've called him that since we were kids. He's just so smooth—like skin. Did you know he used to be a jazz musician?" She laughed as she remembered the origin of the nickname. "He used to cut his hair short when everyone else had afros. You could see his skin."

Rev. Theodore has a playful yet quiet, gentle spirit, but it seemed as though he had summoned the power of God Almighty with his flute that day. The strength of the music alone, without the lyrics, was so powerful I could hear the congregation weeping. Rev. Theodore's solo was aptly named "The Hymn of Preparation" in the church program.

Then Rev. Rosemary walked up to the podium and began her sermon with the statement, "We are saved by Grace." Her sermon was based on the day's reading from Exodus 32:1–14. "Moses interfered on behalf of his people and they were saved by grace. Grace is unmerited, undeserved, unasked for. It is a gift grounded in faith. By grace you have been saved, through faith.

"The Israelites did not expect Moses to be gone so long. After consulting with Aaron, they melted the gold and made a bull, a god. This reflected the sin of Israel and violated the law of God. Their worship was nothing more than noise. The words were directed to the wrong god."

She referred to the passage where God says to Moses, "I have seen this people, how stiff-necked they are. Now let me alone, so that my wrath may burn hot against them and I may consume them; and of you I will make a great nation."

"Stiff-necked people think more highly of themselves than others," Rev. Rosemary said. "They rationalize the word of God and make it fit situations. Moses was an unselfish man without his own goals or personal ambition. He was the right man at the right time for Israel. He was the only one who could appeal the destruction of Israel to God."

I had never quite thought of this famous biblical passage in terms of Moses being an advocate for his people. He had listened to God and led them out of slavery, but now things had gone badly. Moses could have replied, "You're right, God! They're worthless. I could not agree with you more. After all you've done for them, just look at them. They deserve to die. Now, when can we start with my new great nation?"

But Moses interceded and pleaded the case of his people. And in the final verse, God spared Israel. Rev. Rosemary explained how prayer had saved a nation.

"Prayer becomes a change agent. God wants a meaningful relationship with his people. Prayer gives emphasis and affirmation to the existence of God. Moses knew the power of interceding prayer."

Rev. Rosemary and Rev. Theodore are agents of change in their community. Rev. Rosemary not only teaches what she calls the characteristics of God—love, compassion, mercy—but she also embodies those characteristics. She intercedes daily on behalf of her community with prayer and action.

Like the church volunteers who cooked what was left of the chicken in expectation of more, Rev. Rosemary and Rev. Theodore kept tending to their community day after day. They drove people who no longer had cars to doctor appointments. They picked up church members who had been evacuated to other hospitals in neighboring states. They bought groceries for the Children of the Storm. Mt. Zion was the community's Mt. Sinai. It was a place where great men and women interceded in the aftermath of a storm that would have destroyed the spirit, hope, and love of their community, had they not acted with love, compassion, and mercy.

Ms. Bowser walked over to me after the service. She knew I had been trying to help people get federal relief aid. She joked with me and whispered, "Can you make my FEMA trailer BIG?" Then she put her hand over her mouth and ran away as though she had just said something naughty. I laughed and she came back, leaned close to me and said, "Well, I guess this ain't Burger King, is it?" You can't "have it your way" when it comes to FEMA trailers.

Our laughing was interrupted by the sound of singing. A toddler had made her way to the microphone and was singing a solo. The little girl was Ms. Bowser's granddaughter, Rhayna. Ms. Bowser ran over to help her adjust the microphone. When she was finished with that song, the two of them sang a duet, "This little light of mine, I'm gonna let it shine." Ms. Bowser's pain and frustration had melted away under the warmth of the Mt. Zion community and the glow of a grandmother's pride.

Ms. Bowser and the Children of the Storm had been transformed by grace. The love that abides within the community of Mt. Zion had transcended their pain and suffering. It had broken down my own walls of anger and erected a temple of hope—hope founded on the redeeming power of love.

The depth of my commitment to love unconditionally was tested a few weeks after my second trip to Mississippi. Ellen e-mailed me a copy of a November 1, 2005, *Washington Post* article titled "Lesbian Minister Defrocked by United Methodist Church." The first sentence of the article reads, "The highest court in the United Methodist Church yesterday defrocked a lesbian minister in Philadelphia and reinstated a Virginia pastor who had been suspended for denying a gay man membership in his congregation." Both Mt. Zion and St. Paul are United Methodist churches.

To say that this decision did not hurt me would be a lie. It was painful to see the United Methodist Church doors slammed shut to an entire community—my community—when I had seen the overpowering, nondiscriminatory, and unconditional love of two of their churches in the wake of a human disaster.

The ruling helped me understand how Ellen had felt as she was sitting in the Roosevelt Room as part of the White House Press Corps when President Bush announced that he would support a Constitutional Amendment to ban her from being able to declare her commitment to love another in marriage before God and government.

While I do not know the positions of Mt. Zion or St. Paul as congregations, or of Rev. Rosemary or Rev. Theodore as individuals, on the United Methodist Church's ruling—or on me—I remain committed to serve them, just as I have seen them serve their *entire* community. The essence of unconditional love is to be able to love and serve even those who think that I am unworthy of their love. As my brother Chip teaches, "If you have to believe like me in order for me to serve you, then I am not a servant."

Anger and bitterness are as destructive to the human spirit as violence. I aspire to remain true to what I believe is the heart of God—unconditional love. I aspire to be like Rev. Dr. Martin Luther King Jr., who had the capacity to love even in the face of lethal discrimination and hate. "I've seen too much hate to want to hate," King said. "I say to myself hate is too great a burden to bear. Somehow we must be able to stand up before our most bitter opponents and say: 'We shall match your capacity to inflict suffering by our capacity to endure suffering. . . . Do to us what you will and we will still love you.'"

ten

OWNERSHIP SOCIETY

"Up until now, I've tried to be the best preacher/
pastor I could be, teaching the Word of God and
loving the people God has given me to lead.
However, Katrina seems to be giving me a new look
at my job. Anything I can do to help our people, I
am willing to do. Anything you can do to help us
will be of great value. Thank you for caring."

—*Sonny Adolph, Pastor, First Missionary*
Baptist Church, Gulfport, Mississippi

Shantrell introduced me to Rev. Adolph. He wrote this to me in
an e-mail six weeks after the storm; the time stamp on the message was after midnight. Rev. Adolph, like Rev. Rosemary and
Rev. Theodore, had been thrust into a new ministry—a ministry of home building and home repair.

Until you've spent time with people who have lost everything, including their homes, you can't fully understand the
importance of home ownership. Many of Katrina's survivors
had saved their whole lives to buy a place to call home. All of
the people we met owned their homes (the renters had no
choice but to leave town).

They labor ten to twelve hours a day doing work that most of
us could do for only a few days at the most. Although many are
skilled and educated and have phenomenal emotional intelligence as well as common sense, they are unlikely to rise above

"just getting by," but they *love* their homes. No matter what the size, their homes are their pride and joy.

Owning a home marks a solid line in the socioeconomic continuum of life in America: a line between being in the chronically poor lower class and being in the middle class.

The distinction comes down to the ability to borrow money. Banks loan money to the people who they think will repay them. They typically approve loans for people who have steady jobs, some savings, and have demonstrated financial responsibility in the past. According to the U.S. Census Bureau, 69 percent of Americans were homeowners in 2005. Fewer than half of Americans owned their own homes in 1900. In Mississippi, 72 percent own their homes. At 61 percent, Mississippi ties South Carolina for the state with the highest percentage of black homeownership.

Harrison County includes the towns of Gulfport, Pass Christian, Biloxi, and DeLisle. County officials estimate that one-quarter to one-third of the county's 185,000 residents are now homeless. Some, particularly those who were retired or near retirement, owned their homes outright. They were set for life—or so they thought.

As it turned out, Hurricane Katrina inverted the ownership principle: those who did not own property are now better off than those who did. The renters had relatively little invested, so they could leave for other opportunities, and most did.

After talking with Rev. Rosemary and Shantrell, we decided that the most valuable thing we could do for people was to buy material and help them repair their homes. Ellen went with Shantrell and Belinda to buy additional supplies from the lists they had collected in the Gulfport Turnkey development, and Jason and I took the U-Haul to the lumberyards to buy building materials. We delivered our first truckload to the small makeshift supply depot at the Lockett Williams Mortuary. As we pulled into the parking lot, a petite, fit-looking, middle-aged black woman named Ms. Dot came out to greet us. Ms. Dot quickly walked over to us. "Come over here and let me give you a hug. I cannot tell you how much you all have blessed us." She then introduced us to her son, Al, who looked strong enough to single-handedly unload the U-Haul.

We unloaded the two chain saws that we had bought and then decided we needed to find a dry place for the building supplies because rain was forecast for that evening. Al said his place was dry and secure and we could store things in his carport.

Al is a longshoreman. He later showed us one of the barges he had loaded right before the storm hit. We could see from a distance where it had washed up along the shore. Al was a man of few words and was not interested in complaining about his situation. When he told me that he was a longshoreman I said, "I understand that pays well."

"Yeah, it did," he said nostalgically. "I was making twenty an hour. Now, who knows?"

Al's quiet but worried voice was similar to the voices of the men and women I worked with at United Airlines in the wake of a bankruptcy that had lasted almost three years. Most of the employees resignedly accepted the pay cuts. They downsized their homes, they cut their living expenses, and their spouses took on added financial responsibility. The typical expression was, "What are you going to do? It is what it is."

The thing that caused the most anxiety for my fellow employees and me was the unknown—the constant wondering about whether our company would make it. We were hanging on the edge. Another terrorist attack, a health epidemic like SARS, or the decision of one bankruptcy judge could be the end of United Airlines.

Al's anxiety over his job had to be ten times worse than mine. He was his family's sole breadwinner, he had lost his job, and he did not know whether it was ever coming back. To make matters worse, his wife and premature twins were staying with his relatives in Atlanta. The babies were all right, but it was not safe for them to be in Mississippi.

Nevertheless, Al spent the day with us helping his neighbors rather than worrying about his own problems. He and his mother led us all over town in his immaculate green pickup while we drove the U-Haul back and forth between Home Depot and Lowe's and bought building supplies.

The extent of the community's network was impressive. Jason, Al, Ms. Dot, and I were all at Lowe's waiting for a dock loader when out of the blue a woman named Jeannie showed up with a flatbed truck. It seemed Ms. Dot had called her when she heard Jason voice concern over the U-Haul weight restriction. Jeannie and a helper ran over to carry the supplies.

Jeannie is a freelance builder. I overheard her telling the Lowe's dock loader that she lost two relatives in the storm. One was a two-year-old niece. "What you gonna do after that? We have to pick up and move forward," she said. A gentleman showed up to help. Shelton Watts, a fit man in his sixties, looked very tired, but he waited patiently while we chose the proper supplies and helped us load and unload them.

That evening, Shantrell took us to Mr. Watts's home to deliver a gift certificate for building materials. As we approached the door, his young grandson yelled into the dark, "Well, y'all get in here. Come on in now! Hugs, hugs for EVERY-BODY!" He and his sister hugged us, then ran into the next room and came out with toys for Ellen and me. They wanted to give us a gift. Ellen got Road Runner and I got Wile E. Coyote. They were living with Mr. and Mrs. Watts because their home had been destroyed. Their father was still at work. There were pictures on the wall of the Watts's son, who had returned from military service in Iraq just prior to Katrina.

One month later, I received a call from Shantrell.

"I have some bad news for you. Remember that couple we visited with their grandkids?"

"Yes, of course," I answered. "He helped us out with the supplies. His son served in Iraq."

"Yes, that's right. Well, I just found out that he died this week." He had come in from washing the car and fell in the hallway. His son was home and told Mr. Shelton's grandchildren to run to the neighbor's for help. A week later Mr. Shelton died in the hospital. He had no known health condition.

I choked back my emotion as I thought about his two beautiful grandchildren who had been so generous to us. They had lost their home, their clothes, and their toys, yet they wanted to give something to us. I recalled watching their faces light up as Ellen and I played with our Road Runner and Wile E. Coyote. They had so much joy and enthusiasm for life.

I looked back to that night when we arrived at their house. Everyone was ready for bed and the couch had been pulled out and made up for the children. Perhaps their enthusiasm had to do with the fact that the visitors were delaying their bedtime, or perhaps it emanated from the unique bond of love that exists between grandparents and grandchildren. Katrina took the children's home and now she had taken their grandfather.

A United Airlines captain I met on my last trip before going to Mississippi explained that the stress caused by a hurricane could kill a person. The captain was catching a ride with us to his home in Orlando, Florida. I told him that I was going to Mississippi to deliver relief supplies and asked if he had any advice based on his experiences with the hurricanes in Florida.

"I don't know what to tell those people. Those hurricanes just about killed me," he confided.

"You mean you rode them out?"

"Hell no. It was the stress of the aftermath that got to me. I told my wife that I didn't think I could survive another one."

Ellen, Jason, Adam, and I knew from the first time we stepped into a Gulf Coast home that the whole region was on the verge of a health crisis. If the stress didn't kill them, the mold might. Sinus infections, nausea, diarrhea, and vomiting were starting to inflict the oldest, youngest, and most stressed.

For those lucky enough to have some semblance of a home standing and a blue tarp on their roof, the primary threat was, and still is, mold. Left untreated, mold will take over an entire home within a few weeks. There was neither time nor resources to treat the mold, and the mold won. Most homes need to be gutted, stripped down to their frames, scrubbed with bleach, dried for days, and rebuilt.

We were eager to get to work. Never having torn down drywall and insulation, I quickly gained an appreciation for the men and women who do this kind of work for a living. For days afterward, my arms itched from the fiberglass insulation. I learned the hard way that the more you scratch, the deeper the fiberglass shards penetrate the skin. It would have made sense to wear long-sleeved shirts, but we didn't have any and even if we had, it was too hot to wear them. The temperature outside was in the high nineties with humidity that brought the heat index to over a hundred degrees.

We began our demolition and rebuilding efforts on the home of Ms. Verlon, whom I called "Ms. V." I relearned my manners while in Mississippi. People who are senior in age are called ma'am or sir, and their first or last names are prefaced with Ms.

or Mr. Ms. V is an attractive middle-aged woman, but she was exhausted from battling a sinus infection for several days.

Ms. V's oldest son, Anthony, is one of Shantrell's best friends. They have known each other since the second grade. Anthony is a vice principal who works with Shantrell's husband, Myrick, at Gautier High School.

Ms. V is also a single mother. When Anthony was young she earned a slot in one of the first government work-training programs offered in the Gulf Coast and qualified for a job that enabled her to buy her own home.

Ms. V is a marketing director for the Gulfport Convention Center. FEMA had offered her shelter six hours away, but she did not want to leave her home for fear she would lose her job. Ms. V., her two younger sons, Leroy and Josh, and Leroy's two-year-old son rode out the storm in her home. The rainwater had broken through the roof, causing the ceiling to buckle. They listened to the radio throughout the storm and were saved by the advice they received telling them to poke holes in the ceiling to let the water drain and avoid a complete ceiling collapse.

The water broke through large pieces of ceiling anyway, but no one was hurt. All of her ceilings were still sagging. When we tore down the sheet rock, a whole side fell down and almost took out our little work crew. We started out thinking that only two rooms were damaged, but the more sheetrock we took down, the more black mold we found.

We did as much work as we could until dark. I felt terrible leaving Ms. V and her home in this condition. Jason promised to stay on and at least replace the drywall, and Adam delayed his trip back to Washington, D.C. As Ellen and I said good-bye, Ms. V thanked us and said how proud she was of her new home (even though it only had one intact room).

"Come on, Mama, you can't stay here any more," Anthony said. She assured him that she would go over to her sister's house. He knew she was placating him. "Mama, this mold has made you sick. Now come on." Anthony couldn't convince his mother to go to his house. She was too proud and too independent.

Jason became Ms. V's general contractor and trained her sons and a worker from Anthony's church in the art of home construction. Myrick spent his evenings at Ms. V's, after working all day getting the school prepared for the students.

Katrina left homeowners of all ages and races bereft. A woman at the Harrison County Supervisor's Office, named Ms. Judy, had been a school counselor for three decades before working for the county. She is nearing retirement from her second career. Due to the storm, Ms. Judy and her husband had fourteen family members living under their roof.

"Oh, we're all right," she said, "but I just know that those walls are going to have to come down. They're soakin' wet with water."

"Well," she said with that dry, Southern sense of humor, "I suppose I can work till I just drop dead."

Ms. Judy reminded me of my own mother. Her looks and demeanor are very similar: perfectly dressed and made up, organized to a fault, did not blame anyone for anything, and refused to be a victim or to let the circumstances get to her. I later learned how Ms. Judy might have developed those character traits; she has Lupus, a rare, debilitating, and potentially life-threatening disease.

Ms. Judy has worked nearly her whole life. Her husband is working seven days a week for the power company. They are nearing retirement. They had some savings, but not enough to rebuild their home and start over.

After commenting that she didn't think their hurricane insurance was going to cover much (she was told to expect about $12,500), she surprised me by asking, "What do you think we should do?" I was used to a one-way line of questioning. I instantly inventoried my mind for options. Since my first trip to Mississippi, I had spent hours and hours researching available aid and canvassing every federal and state agency that had a hand in the recovery effort. Nothing came to mind. Ms. Judy and her husband had done everything right. They had worked to buy a home, they had planned their retirement and had saved for it. Their investment was insured. Now everything they worked forty years to build needed to be torn down and rebuilt, but they didn't have another forty years in which to do it.

"Ma'am," I finally offered, "the only thing I can think of is for you to apply for a Small Business Administration (SBA) low-interest loan. You may be able to save a few percentage points on the loan."

"Oh, I don't want a handout," she shot back.

"Yes, ma'am, I understand that, but you are in a unique situation here. It really is okay to take advantage of this opportunity. You are not going to take away from someone else."

She smiled and thanked me. I doubt that Ms. Judy ever applied for an SBA loan.

I had a chance meeting with a couple in a similar situation. I was at the only open coffee shop on the entire Gulf Coast, PJ's Tea and Coffee Company. The couple had a stack of pictures along with receipts, financial records, and a blue form that read, "SBA Loan Application." I sat at the table next to theirs and worked on my computer as they pored over every detail of the form and cross-referenced their records. I felt a little awkward sitting so close while they discussed the most intimate details of their household finances.

As they were gathering their things to leave, I asked if they were filling out an SBA loan application.

"Well, tryin' to," the woman said. "You practically have to be a genius to do it." Come to find out, this woman may have been a genius: Ms. Libby Graves has a PhD and teaches high school physics, chemistry, and environmental science. Her husband is a school psychologist.

They had sunk their life savings into a rental property near the water. They had the before, after, and after-again pictures of what was to be their dream retirement home. One of the after-again pictures was of a coffin in the little yard that had washed up from the mortuary located a block away. She assured me that the coffin was vacant when it got to their yard.

Their primary home, also in ruins, was several miles from their rental property. They were living in their van. They could run inside their home long enough to take a shower, but because of the mold, they couldn't linger any longer. They were still teaching full-time. Their future was riding on that SBA loan approval.

We talked some more and then two women joined our conversation. It turned out that they'd all known each other when they were young but had not seen each other in years. One of the women, Ms. Mary Ann Mach, almost did not survive the storm. She looked like she weighed about ninety pounds soaking wet and was probably in her seventies. She watched the tidal wave from her home through a little peephole in a boarded-up window. She grabbed her dog, Cassie, and ran to the attic. She described the force of the water that crushed her home below. "You know, my father's sister had a grand piano and all I found left of it were three casters. They were flattened like pennies."

Ms. Mary Ann reminded me that Southerners treasure their

family heirlooms as much as their homes. I could see the sadness in her eyes as she contemplated the loss of her family's history. And then suddenly her face lit up. "You know what? I'm gonna make myself a little shadow box out of those casters."

A few days later I found myself helping people fill out the same SBA loan applications that Ms. Graves had jokingly said required a genius IQ. Rev. Rosemary had asked me in front of the congregation at Mt. Zion's Sunday service if I would be available to help people with the loan applications one night at the church. It was a request I could not refuse—even though I had been intimidated just by watching Mr. and Ms. Graves attempt to fill it out.

Fortunately, Shantrell agreed to help me. When we got to the church there were already about ten families waiting in line for help. One couple's home had been totally paid off, but it was destroyed. They said they had no outstanding debt, not even a car payment. They had $18,543 in savings. He had already retired and his wife was due to retire soon with a teachers' pension. The loan application asked two critical questions they could not answer:

1. How much insurance money or FEMA money or any other source of financial assistance did you receive, or will you receive? (The form said to leave it blank if the answer was unknown.)
2. How much will it cost to replace your home? (Again, instructions said to leave it blank if the answer was not known. I could not imagine how the SBA could determine how much money to lend if they did not know how much money an applicant had and/or needed to have to rebuild.)

"We just don't know the answer to either," the man applying for the loan said.

"Who knows about insurance?" his wife said. "We've heard as little as maybe $20,000. It could be over $100,000 to rebuild. We have no idea."

As we were finishing the application I asked, "Now you're sure that you don't have any other debt? No car payment, nothing?" As they thought about it, I could see a glimmer of pride on their faces as they shook their heads no. And then suddenly, almost simultaneously, their expressions changed to dread. "Oh no, I forgot. Our truck was destroyed so we had to replace it. Yes, we've got a note on that truck," the husband said.

"We haven't even sent in the first payment," his wife added. She drew in a deep breath as she contemplated how long they would be paying for that truck.

Another woman I attempted to assist is a home care provider for the elderly in the community. Her husband is disabled and on a fixed income of about $600 a month. She inherited her home from her parents and borrowed approximately $12,000 to add a room, install a new roof, and replace the floors. She showed me the water-stained loan payment book as well as all the cancelled stubs she had as proof of payment. She had been paying the debt faithfully for over a year at $350 a month, but she still owed almost the entire principal. I smelled a loan shark.

She was still working, but her car was destroyed by the hurricane. She was forced to rely on others to get to work and that meant she was working about 30 percent less. Her teenage sons were waiting in the church as we discussed her options. She had $354 in savings and she did not owe any other money except her home improvement loan. There had been four feet of water in her home, so I knew it was going to need to be rebuilt.

It took us a few hours to get through all the applications. People asked me if I thought these loans would come through. I didn't know—it was so hard to predict. According to a *USA Today* report on the front page of their Thanksgiving weekend edition, more than eleven weeks after the storm the Small Business Administration had approved fewer than 8000 of the 205,000 loan applications.

I read the congressional testimony of SBA administrator Hector Barreto, explaining away the poor approval ratings. He reminded the legislators that people still had to meet the standard loan qualification requirements and that many would not get approved as a result; but they had to go through this process in order to be eligible for a grant. The SBA loan application process seemed like another bureaucratic exercise in futility—expectations are raised, hoops are jumped through, and then comes the inevitable rejection.

After working through several SBA loan applications, I tossed and turned all night, dreaming about loan applications, mold, and sheetrock. I also had one of my recurring anxiety dreams—the one where I'm trying to get to the airport to fly a trip, but a thousand different impediments delay me. The stress ultimately wakes me up before I ever get to the airport.

While I could wake up from my nightmare and go home to New York the next day, the people of DeLisle and the Pass could not. They *were* home. They had struggled to buy those homes and they would have to struggle again to save them.

eleven

A DATE WITH FOX NEWS IN NEW ORLEANS

The media images that first inspired me to go to the Gulf Coast were not of Mississippi. In fact, the only image of Mississippi I remember seeing until I got there was of Governor Haley Barbour on cable news. The only time we saw television cameras in Mississippi was at the Holiday Inn swimming pool/ dolphin aquarium. The "story," as far as most of the media was concerned, was in New Orleans.

Once we arrived in Mississippi, I had no desire to go to New Orleans. We had plenty of work to do where we were, and Mississippi was not getting enough attention. We could help. Ellen agreed, but she needed to find a satellite feed to do her weekly Fox News Channel segment, *The Long and Short of It*. Jim Pinkerton is the "Long" part of the show, standing 6 feet 9 inches tall, and Ellen is the "Short" of it at 4 feet 10 inches. Jim is the conservative and Ellen is the liberal. Naturally, the only place that had a satellite feed remotely close by was New Orleans. Fox News Channel, along with every other news organization in the country, had a presence in New Orleans.

Ellen called the Fox producer in New Orleans to get detailed instructions. He knew that there were no signs, so he described the route with mileage distances and visual landmarks as best he could. He said if we could find the Superdome, we could find Fox, whose makeshift studio was nearby. We mapped out the plan the night before. It's normally a one-hour drive from Gulfport to New Orleans on Interstate 10, but Katrina had taken a big chunk out of the highway, so the drive took almost three hours. We left DeLisle at 3:20 AM.

The journey reminded me of driving in Iraq with the Marines. We were dodging debris everywhere in complete darkness, only now we were in a Chevrolet rental car instead of a Humvee. Without street signs, we were never exactly sure we were going in the right direction. I worried that we would get lost.

On the outskirts of New Orleans, we came to a flooded underpass. "He (the producer) said to expect flooding," Ellen announced. We knew we were in New Orleans because we had passed a military checkpoint, but we were not sure where we were or whether we were headed toward the Superdome. We couldn't see the Superdome, our major landmark. Fox was supposed to be about a mile and a half from the Superdome. I stopped short of the dark pool of water and Ellen said, "Gun it."

"I'm not going to gun it. First we need to find out if we're on the right road."

"He said to expect flooding," she repeated.

"I know, but who's going to pull us out of here if we get stuck?"

Pilots are taught a useful problem-solving technique when it comes to emergencies: first, maintain aircraft control. No points are awarded if you lose control of the airplane while fixating on some problem inside the cockpit. Second, analyze the situation. Use all the resources available to you, in and out of the cockpit, to help accomplish this step. Third, take the appropriate action: if you need to apply a checklist, shut down an engine or some other procedure, do so. Finally, land as soon as conditions permit.

We needed to move through these four steps quickly or Ellen would miss her Fox News hit and I would most likely suffer the consequences. I stopped the car.

"You get Fox on the phone and find out if we're heading in the right direction. I'll check out the depth of the water."

While Ellen attempted to reach the Fox producer, I made some waders out of two plastic shopping bags. They almost covered my knees. I was quite pleased with my ingenuity until I put them to the test. My feet and legs were soaking wet with New Orleans' toxic black water. Mental note to self: *Shopping bags provide zippo water protection.* I had to laugh at myself for being such an idiot. Since I was already contaminated, I might as well wade in farther. The water was over my bare knees, but the ground had leveled out and I decided that we could probably make it.

By the time I returned from my toxic wade, Ellen had gotten in touch with Fox and was pretty sure that we were headed in the right direction. I gunned it.

Our little emergency lasted about ten minutes. Still no Superdome in sight. We came upon a rare, clearly marked inter-section—the signs were bent, but they existed and were read-able. We were at the corner of New Orleans and Humanity streets. It was perfectly dark and still. There wasn't a soul around. I stopped to take a quick picture. I wanted to remember this moment.

We drove a few more miles and finally caught our first glimpse of the Superdome. I was relieved and sad at the same time: relieved that we weren't lost, sad because the Superdome stood as a monument of suffering.

We still had to find Fox. Ellen's cell phone was cutting in and out, until suddenly she had a clear connection with a retired police officer from Atlanta who was acting as the Fox News security guard. He gave us point-to-point instructions to the makeshift newsroom and satellite feed. We got to Fox with about thirty minutes to spare—just in time for Ellen to put on her make-up.

After Ellen's Fox News hit, Rob Milford, a radio broadcaster whom I met during the war, saw Ellen and said, "Where's that

U-2 pilot you hang out with?" Ellen brought him over to me as I searched for coffee.

"They'll let anyone into this town!" Rob said.

"He's going to take us on a tour!" Ellen announced. I was relieved that Rob was willing to be our tour guide. While I love an adventure as much as the next person, this was a bad time to wander around the streets of New Orleans. Pilots are trained to minimize risk, not look for it.

Rob was the perfect guide. He was covering Katrina for Fox News Radio. "Radio Rob," as I call him, is a fifty-something soldier of fortune, of sorts. He loves to play war and does it very well. He knows the units, ranks, and general workings of the U.S. military. Rob was right at home in the post-Katrina state of New Orleans.

"The gun's locked and loaded in the yellow plastic shopping bag. You know how to use it if you need to," Radio Rob said with full Alpha Male bravado. We hopped into his rented Chevy Blazer.

New Orleans was under the control of the U.S. military. There were no laws except what the military would or would not allow. The military didn't care about traffic laws. They were there to maintain order, even though there were hardly any people in the city besides recovery-related workers and journalists.

Rob is not afraid of being told "no." Neither does he seem to give a lot of consideration to consequences. He uses his knowledge of military nomenclature to push the limits. This allowed him to whip around New Orleans as if he owned the place. When we would get stopped at a military checkpoint, he would turn on the charm with something like, "Ah, the 82nd (as in 82nd Airborne) is here. Now I know everything's gonna be all right." The soldiers seemed to know they were being manipulated, but then they probably don't run across many people who can identify their rank, unit, and history like Rob can.

The French Quarter looked better than I have ever seen it. There was no trash on the streets and it looked like it had been scrubbed down. Even with a whiff of rotting food in the air, it smelled better than ever; the occasional smell of urine and vomit is the norm. Rob explained that the Quarter had been spared much flooding and that businesses were scheduled to start opening that morning. As he pointed out each journalist on the street, he joked that there were three of them for every French Quarter shop owner.

We stopped and peered through the gates of Jackson Square, where President Bush had addressed the nation just three days before, on September 15, 2005. St. Louis Cathedral radiated a soft glow as the rising sun began to reflect off her facade. She was alone, just as President Bush had been when he stood in front of her and boldly acknowledged our nation's shortcomings and pledged to bring her people home. Shantrell, Ellen, and I listened to his speech as we huddled in close to the car radio that night:

> Our third commitment is this: When communities are rebuilt, they must be even better and stronger than before the storm. Within the Gulf region are some of the most beautiful and historic places in America. As all of us saw on television, there's also some deep, persistent poverty in this region, as well. That poverty has roots in a history of racial discrimination, which cut off generations from the opportunity of America. We have a duty to confront this poverty with bold action. So let us restore all that we have cherished from yesterday, and let us rise above the legacy of inequality. When the streets are rebuilt, there should be many new businesses, including

minority-owned businesses, along those streets. When the houses are rebuilt, more families should own, not rent, those houses. When the regional economy revives, local people should be prepared for the jobs being created.

Americans want the Gulf Coast not just to survive, but to thrive; not just to cope, but to overcome. We want evacuees to come home, for the best of reasons—because they have a real chance at a better life in a place they love.

There was silence afterwards, the silence of new hope. I wanted to believe that our president would keep the promises he had made to this city and all of the people of the Gulf Coast, for this was a campaign of a different sort, a campaign for humanity, not power.

Rob's deep and dramatic radio voice interrupted my thoughts, "Enough of this. Now you really need to see the rest of the story. Hope you two don't have weak stomachs." He drove us to the Ninth Ward. It still resembled a flooded sewer almost three weeks after the storm. You could see by the watermarks on the homes that the standing water had been as high as seven feet. It smelled like raw sewage. The water got so deep Rob's Blazer began to make a wake, so we decided we'd better turn around.

As we drove by the flooded homes of the Ninth Ward, I began to think about the people who owned these homes. These flooded dwellings belonged to the same sorts of people I had been with in DeLisle and the Pass, only they had no chance of coming home any time soon, possibly never.

Stray dogs waded slowly through the toxic waters of the Ninth Ward. One lucky dog had been adopted by two National

Guardsmen and he was keeping them company at their check-point. "All he does is lay here all day, but we're tryin' to save him," they said of their new friend. The military is famous for adopting "critters," as they say in the South. Critters are a wel-come source of unconditional love and affection in worlds that have neither. The platoon I was with in Iraq wanted to adopt a stray Iraqi puppy on the first day of the invasion. One of the men keyed the mic and said in a southern accent, "Lieutenant, can I trade the pigeon for that Iraqi (he pronounced it "Aye Rackie") puppy?"

Each platoon was assigned a pigeon to ride out the war. The pigeon was supposed to be a live Weapons of Mass Destruction litmus test. If it died, then chances were we would die. The Lieutenant denied the request and the pigeon stayed on the job.

New Orleans seemed to be under military occupation. The Humvees, checkpoints, men and women in flak jackets car-rying M16s made me think I was in another country. I could not believe that a city in my own country had so thoroughly col-lapsed under the weight of chaos. The city had been known throughout the world as the grand dame of so many things, including jazz, food, and fun. I used to call her the New York City of the South, not for her pace certainly, but for the diver-sity of her soul and the strength of her voice. All had been wel-come to pass through her gates; now few could survive within them. The self-contained U.S. military remained as her guardians to restore and maintain the most basic elements of any society—law and order. She stood naked, silent and alone, stripped of any title she may have held.

Rob had to get back to work and so did we. As we left town, we drove by the now famous staging area on Interstate 10 where people were first taken after they were rescued, before being evacuated out of the city. We saw the plastic bins that had held the few mementos and clothing the victims tried to salvage. Their belongings were now scattered like trash after a football game. I wondered what had happened to them. What were their lives like now? What would become of them and their hometown?

twelve

DAMAGES NOT COVERED

Ellen's right-wing talk show colleagues are fond of the motto "God helps those who help themselves." Personal responsibility is the antidote to suffering. The people of the Gulf Coast took responsibility for their property; they were heavily insured against hurricane risk. Unfortunately, it looks as though the vast majority of homeowners' damages caused by Katrina will not be covered—even for those who paid hurricane insurance premiums year after year.

As of October 4, 2005, the Property Claim Services (PCS) unit of the Insurance Services Office, Inc., reported that there were 490,000 insurance claims in the state of Mississippi, accounting for approximately 30 percent of the total Katrina claim count between the six states affected (Louisiana, Mississippi, Alabama, Florida, Tennessee, and Georgia).

One would assume that an insurance policy that specifies hurricane coverage would cover damage that results from a hurricane—ergo, hurricane coverage. So why isn't the water damage from Katrina (a hurricane) covered? The reasoning, according to an industry trade group, the Insurance Information Institute, is that "much of the damage from Hurricane Katrina is the result of flooding, rather than wind." The devil's in the definition, so to speak.

Most of the water damage in Harrison County has been written off as flood damage according to the insurance companies that use FEMA's guidelines to define a flood. Unfortunately, less that 12 percent of Harrison County residents had flood insurance. Robert Hartwig, PhD, is senior vice president and chief economist for the Insurance Information Institute. On one page of a 198-page PowerPoint presentation dated November 4, 2005, titled *Hurricane Season of 2005: Impacts on the Property/Casualty Insurance & Reinsurance Industries*, he includes a chart that lists flood coverage across the coastal communities. In Mississippi, less that 12 percent of homeowners were covered in Harrison County, less than 11 percent in Jackson County, and less than 24 percent in Hancock County.

Of course, one would assume that some damage was caused by wind-driven rain that penetrated the home and therefore should be covered. Wrong again. Most insurance policies exclude mold damage even if the water damage that initially caused the mold is covered. Mold coverage is extra.

Hartwig describes the standard mold exclusion in his presentation: "We do not cover loss or damage, no matter how caused, to the property which results directly or indirectly from fungus and mold."

The Gulf Coast is a giant mold incubator. Everyone has mold damage. Even those who thought they were spared later found that they were not. Shantrell's and Myrick's home had wind-driven water damage. They were fortunate that the damage to their home was covered. The insurance adjuster came out, looked it over, and said he would send a check. He did not see nor did he mention mold. A few weeks later, Mason (who was twenty months old) and Taylor (who was three years old) began vomiting and had chronic diarrhea. Their home had become infested with mold.

They had to move out and have their walls torn down eight weeks after the storm. The frame had to be treated and dried with a giant machine that continually tripped the circuit breaker due to its power draw. The house had to be reconstructed from scratch. So even with relatively minor damage compared with the rest of the community, the mold eventually prevailed.

Because she is an attorney, Shantrell has an eye for detail. She carefully read her insurance document and discovered the mold exclusion clause, and on her own initiative paid additional money for mold coverage. Yet more than five months after the storm, Shantrell and Myrick had not received compensation for their losses. Shantrell has made dozens of calls to her insurance company in vain.

No one stood a chance against the mold. People did not have the resources to treat the mold—either by tearing down and rebuilding or by attempting to dry out the wet walls with machines that were supposed to fall from heaven and operate with no electricity in the aftermath of the storm. In addition, radio announcements advised people to wait for their insurance adjuster and/or FEMA inspector before they completed major home repairs. I would call ripping out walls major repairs.

I drafted the following question regarding mold to FEMA:

> People were told not to do anything until FEMA and/or their insurance company looked at the property, and now they have mold. Also, there was no way to treat the mold without electricity; mold is as destructive as hurricane force winds. What is FEMA's plan to address the rebuilding costs associated with mold?

Mr. Butch Kinerney of Department of Homeland Security (DHS)/FEMA Public Affairs answered in the following e-mail:

> FEMA never advised anyone to wait on cleaning up mold/mildew. Here is a guide to cleaning up

mold: http://www.fema.gov/pdf/reg-x/mold_mil dew.pdf. Obviously, since electricity has not been restored by power companies in some areas, it will be harder to dry out buildings and mold will continue to be a problem, but there are steps anyone can take: eliminating the source of moisture, opening windows/doors to dry out areas and begin cleanup with the bleach solution.

After reading Mr. Kinerney's answer, I wondered if he had ever tried to "scrub" black mold out of water-soaked sheetrock. It is impossible. I am also unclear on how people without phone service or computers could have gone online to download FEMA's mold guide. Mr. Kinerney continued:

FEMA is not, and was never designed to be, a substitute for proper insurance coverage. As with any damage, insurance is always the first-line of defense against nature. FEMA provides, as detailed above and throughout our web-site where all of this information resides (www.fema.gov), for immediate needs following a disaster and some limited help for those who are uninsured. Additional federal assistance is available through the SBA loan program and other federal programs.

His answers made me feel like I was navigating through a cir-cular fun house where the floor keeps shifting. Here is what he was really saying: We know there was a hurricane, but most of the damage isn't covered by hurricane insurance. FEMA covers damages when there is no insurance, but FEMA is not a replacement for insurance.

I made my second trip to Mississippi six weeks after the storm; people were just starting to see their insurance adjusters. Seventy-five days after the storm, some of Rev. Rosemary's congregation still had not seen their insurance adjusters. When Ellen returned to Mississippi in the third week of November, there was still no progress. As of the end of January, most Mississippians were either still awaiting word from their insurance companies or contesting the payout.

One woman whose home is in the Turnkey development had hurricane insurance but no flood insurance, because she did not live in a flood zone, and yet flood waters rose to four feet in her home. She received two checks from her insurance company two months after the storm totaling only $13,000—even after she contested the first settlement. Her home was completely uninhabitable. Everything had to be stripped to the frame and rebuilt with new cabinets, new floors, a new roof, new appliances, and new furniture. She naively thought her hurricane insurance would cover her for hurricane damage, as it had for other people in other states who had survived other hurricanes. It did not.

Those I encountered who received any payment from their insurance companies simply received a check without an explanation of what was covered or not covered. People were afraid that the act of cashing the check would indicate that they waived their right to contest the settlement. Most people I spoke with received between $10,000 and $14,000.

The next line of argument: These people should have had flood insurance and it sets a bad precedent to take care of them.

White House Budget Director Joshua Bolten, in an interview with the *Wall Street Journal* published on September 26, 2005, summed up this position: "About half the homeowners in the Katrina-ravaged flood area had federal flood insurance, and the government will make good on those claims even as it looks for ways to make the program 'actuarially sound.' It undermines the purpose of an insurance scheme for the government to make payments to those who didn't buy flood insurance. If the government becomes the insurer of last resort even when people don't get insurance, then people won't buy any insurance."

Unfortunately, Mr. Bolten is misinformed on flood insurance in the Gulf Coast. Approximately half of the people throughout the greater New Orleans area had flood insurance, but less than 12 percent had it in Harrison County—because few of them were living in a floodplain.

The residents and government of Harrison County learned their lessons about storms after Camille. The post-Camille building codes required structures to withstand 130-mile-an-hour winds. Tropical storms are upgraded to hurricane status when their winds exceed 73 miles an hour, so there was some leeway.

Those Camille-inspired building codes worked well for the homes that were not on the beach. From the outside, these homes looked fine other than a little roof damage. Unfortunately, the source of devastation was the tidal surge and flooding caused by Katrina's wind and rain. No one—not the government, not the private insurers, not the lenders, and certainly not the citizens of Harrison County—had foreseen that a hurricane could cause this degree of flooding. The risk for flooding was so low that very few Harrison County residents and businesses were required to purchase flood insurance.

Those who did not buy flood insurance based their decision on the projected risk—as determined by the federal government. As my mother said concerning the decision to opt out of purchasing mold insurance (she lives in Odessa, Texas, where it's so dry she has to make her own lotion to keep from shriveling like a prune), "People buy insurance to protect against the *probability* of risk." The people of Mississippi bought insurance to protect themselves against the probability of risk—the risk of a hurricane, not a flood.

Legal actions have been filed against the insurance companies. According to his press release, dated September 15, 2005, Mississippi Attorney General Jim Hood has filed a "motion for temporary restraining order against the insurance industry to protect Mississippi's victims of Hurricane Katrina." He states that the provisions at issue "attempt to exclude from coverage loss or damage caused directly or indirectly by water, whether or not driven by wind." This case is not expected to survive in federal court based on FEMA's definition of a flood.

Mr. Hood's legal action does not have the support of other state officials. Mississippi Department of Insurance Commissioner George Dale was quoted in the local Gulf Coast paper, the *Daily Journal*, in November 2005: "I'm in a no-win situation. The compassionate part of me wants everybody's claim to be paid. A reading of policy terms in many instances, however, clearly states that if the loss results from water damage, the standard homeowner's policy doesn't cover it."

By federal law, insurance companies are regulated by the states. Commissioner Dale has said that he also has a responsibility to ensure that insurance companies don't abandon Mississippi post-Katrina. I attempted to buy insurance for a piece of property in Mississippi from my insurance company, (one of the several national companies listed in Mr. Hood's restraining order). The home I wished to insure had less than $5,000 worth of damage and no flooding. I was initially denied all coverage for any peril, including fire. I appealed the decision, but was told there was no longer hurricane coverage available for that region. They referred me to what Shantrell called an insurance loan shark.

What the White House's Joshua Bolten called an insurance scheme perhaps should be renamed an insurance scam. Without regulatory enforcement, the insurance accountability battle will be left to the efforts of private attorneys.

Richard Scruggs, the Mississippi attorney who helped win the $246 billion tobacco industry settlement and Senator Trent Lott's brother-in-law, has also taken up the cause against the insurance companies. However, Mississippi's tort reform laws do not allow class action lawsuits to be filed, so he is taking on individual cases. Mr. Scruggs is attempting to show that the insurers intentionally misled the insured. He also claims that they are "slow rolling" their clients.

Most of the people I met had a story about being given the runaround by their insurance companies. One woman, a nurse at a local hospital, speculated that the insurance companies were using the goodwill, patience, and hospitable nature of the South to delay, dissuade, and neglect those who paid premiums year after year. She said, "They use all this technical babble and intimidate them into accepting whatever is offered."

In an article in the *Dallas Morning News*, dated November 13, 2005, Mr. Scruggs echoed the nurse's statement: "They do this limbo thing. Let it cool down, let these people get desperate. . . . The insurance companies vacillate on making any commitment at all, hoping it'll die down and the news channel will change subjects. This isn't their first rodeo."

As important as it is to hold companies accountable for their actions, the legal route can take years to resolve. The people of the Gulf Coast don't have years to wait or homes to wait in. Senator Thad Cochran (R-MS), chairman of the all-powerful Appropriations Committee, added a $29 billion supplemental Katrina relief budget package as the senators were rushing out the door to catch their flights home for the Christmas holidays. The amount of $5 billion is earmarked for people who did not live in a floodplain, but local residents are skeptical. They worry that the money, like other Katrina relief, will not trickle down to those in need. While this aid is a start, it does not help the hundreds of thousands who had flood or hurricane insurance and are still unable to collect for damages.

At my six-month flight physical examination in November 2005, my doctor asked me what I had been up to.

"I've been working on hurricane relief in Mississippi."

"Well you don't want to know what I think. You have to understand, I'm a conservative. I think people who live on the coast and on the sides of mountains and in earthquake zones have to take what comes."

He said this as he was taking my blood pressure in his Manhattan office. "Well," I replied, "you know New York City is about five years overdue for a category three to five hurricane, don't you?"

"Don't say that," he said, and we changed the topic to flying.

The National Oceanic Atmospheric Association has tracked category three to five storms from 1851 to 2004. As of the end of 2004, there had been ninety-two major hurricanes from Texas to Maine. Had we followed my doctor's advice, we would not have rebuilt the coastlines of Texas, Louisiana, Mississippi, Alabama, Florida, Georgia, South Carolina, North Carolina, Virginia, New York, Connecticut, Rhode Island, and

Massachusetts—all of which had category three to five storms.

Hurricane Katrina sparked an unprecedented chain of events that caused damages not covered by hurricane insurance. Katrina caused the water to surge, the bay to overflow, the creeks to flood, the barges and shrimp boats to wash up onshore, the contents of homes to travel for blocks, homes to shift off foundations, and levies to break. Usually, hurricanes cause wind damage, not mass flooding. The Insurance Information Institute concedes that Katrina behaved differently than most other hurricanes: "The majority of losses in Florida in 2004 were from wind damage." But as far as our government, state regulators, and insurance companies are concerned, the individual homeowners should have foreseen and protected themselves from an event that is virtually unparalleled in American history.

thirteen

A DREAM SPOKEN

"I have a dream that one day this nation will rise up
and live out the true meaning of its creed. . . . I have
a dream that my four children will not be judged by
the color of their skin, but by the content of their
character. . . . I have a dream . . . that one day. . .
little black boys and black girls will be able to join
hands with little white boys and girls as sisters and
brothers. . . . And when we let freedom ring, when
we let it ring from every village and every hamlet,
from every state and every city . . . we will be able
to join hands and sing in the words of the old
Negro spiritual, Free at last! Free at last! Thank
God Almighty, we are free at last."

—*Martin Luther King Jr., August 28, 1963*

I wanted to believe that Martin Luther King's dream had been
realized, but Katrina's winds cleared the smoke screen of gra-
tuitous political correctness to reveal that while we have come
a long way, the dream still lives unfulfilled. Katrina's survivors
of all races taught me that a dream that is still alive is cause for
hope and a call to action.

I was born four years before the death of Martin Luther King
Jr. I was one of those children he dreamed would one day join
hands with black children. And now, more than forty years later,
Shantrell and I were working hand in hand, with our respective

communities and families, striving to rebuild a community that would let children of all races join hands and dream.

Finally, I truly understood the power of the dream Martin Luther King Jr. dared to speak of. His words ignited the flames of hope, which in turn blazed a new reality for my generation and my world.

Where does the power of a dream—any dream—come from? A dream is distinctly different from a goal. A goal is an instrument of measurement. It can be a goal line or a goal post. Goals are specific. I have set goals all my life.

Piloting requires that I set and achieve specific goals. Flying demands precision. Goal: depart and land on time. Goal: maintain an airspeed, course, and altitude. Goal: land in the "touchdown zone"—on the centerline of the runway—with no crab or drift. The flight ends when we set the parking brake, and we leave with a sense of satisfaction for having accomplished our goals. I am driven to achieve measurable results.

A dream is a reach—a leap beyond the possible. People say, "You must be dreaming!" because the path is not clear, except to the dreamer. A dream requires imagination. A goal requires a plan. A dream comes from a place that does not exist and may never exist. A goal is measurable. A goal can be partially achieved. A dream is all or nothing. To dream takes courage.

Martin Luther King Jr. dared to dream. He dreamed of black and white children holding hands at a time when they could not even attend the same schools. Katrina washed away many of the schools originally built first to segregate and only secondly to educate. By the time Katrina made landfall, Mississippi was still struggling to rid itself of a reputation for racism. I lived in Mississippi when it was still racially divided.

My first assignment, as a brand new Air Force second lieutenant, was Columbus Air Force Base, Columbus, Mississippi.

I was entering "Undergraduate Pilot Training," and I could not wait to get to Mississippi. This was going to be the most important year of my life. If I made it through, I would earn my Air Force wings, become an Air Force pilot, and fulfill my dream of flight. If I did not make it, I would probably become a missile officer in some place that had very few trees and subzero temperatures. Missile officers did not seem very happy. Actually, I only knew one of them very well—my old nemesis at the Air Force Academy, Major Jim.

It was July of 1987. I set out for Mississippi in my 1976 yellow Volkswagen Bug convertible. I remember driving between Memphis and Columbus at around midnight. The thick air was suffocating after having been in the high and dry climates of Colorado and New Mexico for most of my life. I pulled over to take a potty break, but after a cockroach the size of my thumb flew at me, I decided I could hold it. I didn't know cockroaches could fly. I pressed on toward Columbus and looked for a place to spend the night. I chose the Econo Chief Inn due to their flashy advertising ($24.99 a night). I buzzed the doorbell and the clerk who had been sleeping came stumbling out. He looked like a character in a movie. He had several brown spots on his tank top and his arm was littered with various tattoos. He reached for a plug of tobacco and awaited my request.

"Yes, Sir, I'd like a single room for one night."

"No problem, Miss. I'll tell you what. I'll even make sure you don't get a room that a nigger's slept in."

It took me a moment to recover from the shock of his statement. "Sir, that's really not a problem for me. I'll take any room." To this day, I regret that I did not turn around and walk out of there. Why didn't I? Perhaps it was too much trouble to go and find another room, but there was no reason that could have justified my decision. As I tried to go to sleep that night I thought about how grateful I was for the people who did not go along as I had just done, but instead who stood up to racism.

The next day, pilot training began. The pilot training class consisted almost exclusively of classmates from the Air Force Academy. The Air Force scheduled all new academy graduates to begin training between June and August—it was as if an entire civilian college graduating class went to work for the same company. We were happy to reunite and begin *real* life, versus the life we had known at the "Blue Zoo."

Our new commanding officer, whom I'll refer to by his first two names only, walked in and we called the room to attention. (Even though he was only a first lieutenant, we decided to try to make a good first impression.) He told us to sit down and then he began his speech to his new students.

"My name is Lieutenant Jim Bob. I was born and raised in Columbus, Mississippi. Right here. I've lived in Columbus my whole life. Some people say we got a race problem down here. That's just not true. The black folk stay on their side a' town and the white folk stay on their side a' town and you would be best advised to do the same."

It felt as though someone had sucked all the oxygen out of the room. I couldn't believe what I was hearing. I thought that I had taken a wrong turn and ended up in some other planet's Air Force—the planet UR-anus.

This was the antithesis of my first day of basic training as a cadet at the academy. The upperclassman in charge of my flight of about thirty cadets said, "Now listen up, Basics! Wherever you came from means nothing. We will not tolerate any kind of racism here. There is no such thing as black or white or brown or purple. We wear hats so that we can't even distinguish each other by the color of our hair. This is your new family. We are your new family."

That speech made sense to me. I am a "half-breed," half Spanish and half Irish. I was born in Espanola, New Mexico, which is approximately 80 percent Hispanic. Espanola is my hometown. I spent all my summers there as a child and teenager. People there used to call my brother and me "coyotes," the Spanish nickname for half-breed. Since I am a coyote, I never understood racism.

First Lt. Jim Bob's speech seemed like a foreign language. The class took a break after he finished. We all looked at each other thinking that we had just met, as my father was fond of saying, a nutcase. As our commanding officer, this nutcase was in charge of our fate.

I remained in Columbus for four years after completing my pilot training and I trained new pilots to fly. The town's racial divide remained. There were no signs that said "Whites Only," but Lt. Jim Bob's words rang true. Fellow officers were advised by others not to send their children to private schools: they were only used to segregate, and not worth the money.

Most restaurants had either black or white customers; rarely did the groups mix. When they did, it was usually at a very reasonably priced establishment. While it initially occurred to me that the segregation was based more on economics than race, I noticed that even in the integrated eateries, the tendency was for black people to sit on one side of the restaurant and white people on the other.

The military does not overtly attempt to change the social mores of the local, hosting population, but sometimes the presence of the military can be a catalyst for change. Our Wing Commander mandated a base-wide boycott of the local golf course after it refused to let one of his officers of Indian descent play golf. He was dark-skinned but had Caucasian features, making him an enigma to the locals at the golf course:

"the black guy that's a white guy." But as far as anyone could see, he was black and black men were only permitted to carry golf clubs, not swing them. The Colonel took a stand for his officer and for the principles that I believe the military exists to protect and guard. I do not know how the local community reacted or what consequences were brought to bear on him, but his principled stand made me rethink my failure to stand.

Silence in the face of racism is a type of endorsement. I was not willing to confront Lt. Jim Bob's racism on day one of something as important to me as pilot training. Confrontation did not even enter my mind. I just passed it off as Jim Bob being an idiot and went on about my business of learning to fly.

The students at Myrick and Anthony's high school are taught to stand up for what is right, to draw a line in the sand that must not be crossed, even when there are potentially grave consequences for the one taking the stand. This demands immense courage. The sign hanging in the entrance of Gautier High School says, "Courage is not the absence of fear, it is action in the face of fear."

The legacy of courage was revealed to me on my trip to the Mississippi Gulf Coast eighteen years after I had arrived as a second lieutenant. This legacy was embodied in the evidence of real change.

As an attorney, Shantrell shares an office with Brandon Lowther, a successful white oilman who relies on her for legal advice. People of both races trust her with their most difficult decisions.

Gautier High School, located about thirty minutes east of Gulfport, is comprised of over eight hundred students from a mixture of racial and economic backgrounds. Shantrell's husband, Myrick, and her childhood friend, Anthony Herbert, are vice principals. Before coming to Gautier, Anthony, only twenty-eight years old at the time, led his previous school to earn the prestigious Blue Ribbon Award for superior testing scores—given to fewer than three hundred schools in the entire country.

Rev. Rosemary and Rev. Theodore have two daughters—Sonya, who runs the Lockett Williams Mortuary, and Thea, who is the Director of Precollegiate Learner Programs at The College of William and Mary.

Shantrell, Myrick, Anthony, Thea, and Sonya were all young teenagers in Mississippi when I was a jet instructor pilot. They grew up in the South that I knew then. It was better than when their parents struggled for basic civil rights or when their grandparents struggled for basic human rights, but it was still poisoned by an undercurrent of racism.

I have even deeper respect for my hosts in Mississippi because of an experience at United Airlines as a facilitator for a course called "Navigating Change." This course was designed and taught at a time when airlines had money to spend on training beyond the aircraft simulator.

Perhaps based on some related incidents, United Airlines decided that discriminatory attitudes in the cockpit had the potential to adversely impact human performance and, as a result, flight safety. Pilots and company managers worked together with a consultant to design a course targeted at pilots in leadership positions (check pilots, supervisor pilots, instructor pilots, union representatives) in hopes that these leaders would set the tone for the rest of the pilots.

Teaching pilots to think about complex social issues like diversity is akin to teaching cats to swim. We don't want to go there. ("There" being that place where things are not bimodal: good/bad, works/doesn't work, on/off.) We generally steer clear of introspection, nuance, and complexity. Tell me the policy and I'll follow it. I don't need to be bothered with something that isn't going to be tested on a check ride. Yet after the day-and-a-half-long course, predominantly white males between the ages of thirty and fifty-five repeatedly came up to us and said how much the course had changed their views on race, gender, and diversity.

While I would like to believe that my facilitation had something to do with their reaction, the pivotal element for them was the documentary we showed, *Brown Eyes, Blue Eyes*. The day after Martin Luther King, Jr., was assassinated, Jane Elliott, a third grade teacher in Iowa, conducted a class experiment. She divided the all-white class into two groups—those with brown eyes and those with blue eyes. She began by telling the blue-eyed students that they were better and smarter than the slow brown-eyed students.

Elliott filmed the experiment, and the transformation is indisputable. The blue-eyed students turn into confident, high-achieving little monsters, while the brown-eyed students become withdrawn and depressed. Their academic performance plummets. Elliott then switches the roles and makes the brown-eyed students the superior group. Their actions are identical to those of the blue-eyed in the former experiment. Within moments, the children began to live up or down to the label that has been placed on them.

The film *Brown Eyes, Blue Eyes* takes the popular sound bite "the soft bigotry of low expectations," used to advocate universal testing standards, to another level. It illustrates that performance is linked not only to expectations, but also to respect, dignity, equal treatment, and equal opportunity. High expectations mean nothing if the underlying premise is not that a person is an equal first, with equal opportunity.

The children in Elliott's classroom believed that they were *less than* and, as a result, began to perform poorly. Elliott told those children a lie. Both groups, the superior and the inferior, created their own truths from that lie.

Shantrell reminded me of another lie, an age-old myth within the black community: the lighter the color of your skin, the better, smarter, and prettier you are. She explained how what I call the Shades of Black myth still plays havoc in the psyches of those who are born with a darker skin tone. They see themselves as being less than those who have lighter skin.

While the Shades of Black myth sounds crazy to a logical mind, I had to admit to myself that I had bought into my own version of it when I was bumped down from captain to copilot after United Airlines started cutting back on flying following September 11, 2001.

I had been a captain for four months, and when I was notified that I was no longer senior enough to hold the position, I dreaded giving it up.

I didn't appreciate the status distinction between captain and copilot until I was a captain. My world changed the first day I showed up at the airport with the gold braid on my hat and four stripes on my sleeves instead of three. Everyone was super-friendly—even the security workers greeted me with, "Hello, Captain! How are you today?" Suddenly my coworkers asked for my opinion, even on matters unrelated to flying. The flight attendants continually checked on my well-being. I was the same person, but they saw someone *more* when I was a captain than when I was a copilot.

Six months after 9/11, airline cutbacks meant the fourth stripe came off my sleeve and my hat lost the braid on the brim. I was back to being invisible, and I acted accordingly. I focused only on doing my job and went home. Other than trying to fly the plane well, I had no sense of personal pride or connection to my profession. I had bought into my own version of the Shades of Black myth.

Two years later I had climbed the seniority ladder far enough to be a captain again. The tailor sewed the fourth stripe back on, the hat was exchanged for one with a braid on the brim, and once again, I was *visible*.

This experience helped me understand the limitations of the law when it pertains to human behavior, particularly regarding race and gender. Progress has been made in our laws governing the equal treatment of our citizens. "Equal protection" is not merely a political or social cause; it is the law of our land. But laws represent only the lowest common denominator of civilization. Law is a minimum standard. I recalled a quote from an article printed in a libertarian journal: "The measure of civilization in a society is inversely proportional to the number of laws it has."

While overt discrimination is against the public law and the private policies of corporate America, the subtle vein of bias runs just beneath the surface. As long as nothing goes wrong, race is not an issue, but as soon as a problem arises—any problem, the first thing we point to is race. Both blacks and whites point to race as causal in the level of Katrina's devastation and the actions that have followed. Many Americans blame the residents of New Orleans, who were languishing in the Superdome and Convention Center because they did not evacuate when they were told to. Outsiders blamed the black mayor, Ray Nagin, for not getting his citizens out of the city in time. I cannot count the number of times I have heard public and private variations on the theme, "Those people in the Gulf were just generation after generation of black, uneducated welfare recipients."

Others blame much of the suffering on racism. Rap star Kanye West made an off-script assertion during the Katrina fundraising telethon, "Concert for Hurricane Relief," when he said, "George Bush doesn't care about black people."

Louis Farrakhan took the accusation to a deeper level by suggesting a government conspiracy against black people. "There was a twenty-five-foot hole. We've suggested that it [the levee] may have been blown up so that the water would destroy the black part of town."

Jesse Jackson was outraged that people were describing Katrina's survivors as "refugees." He claimed that it implied a certain sub-citizen status, when in fact they were American citizens. Other leaders echoed similar thoughts.

The media loves a fight. It's good business. Soap operas bring in more advertising revenues than documentaries for this reason. I understand the power of language.

One night, after having my fill of Katrina's racial gladiators, I was looking through my Bible when I came across the passage in the book of James that tells of the immense power of the tongue compared with its small size. "Or look at the ships: though they are so large that it takes strong winds to drive them, yet they are guided by a very small rudder wherever the will of the pilot directs. So also the tongue is a small member, yet it boasts of great exploits. How great a forest is set ablaze by a small fire!" (James 3:4, 5) We can build or destroy with our words as well as with our deeds.

These public and private displays of American dysfunction harmed the survivors themselves. The backlash began when people reportedly called to cancel their donations to Katrina survivors after they heard Kanye West's statement that "George Bush doesn't care about black people." The donors focused anger on Kanye West instead of focusing on the needs of the Katrina survivors.

The byproduct of our dysfunction concerning the issue of race is that Katrina survivors—of all races—continue to suffer. The myth that the survivors are just a bunch of entitlement babies who don't deserve handouts justifies inaction. It is easy to justify apathy in the face of suffering when that suffering is borne by people who are not as good as we are. Once the person suffering is devalued, it isn't difficult to stand by and allow the suffering to continue.

I have not met any of these so-called entitlement babies. The Americans I have met carry the mark of a brutal past caused by racism and the mark of a brutal present caused by a hurricane—a hurricane that destroyed the property they had fought for the right to call home. Nevertheless, these Americans have chosen hope and love over fear and anger. It is evident that all of them, generation to generation, have adopted Martin Luther King, Jr.'s dream. Even more impressive than the educational and professional achievement I witnessed among a people who less than forty years ago were not permitted to drink from the same fountain as me, is their genuine love for their entire community—irrespective of race.

It is out of this love that dreams become real.

fourteen

EIGHT EYES, ONE VISION

Rev. Rosemary has a new dream. She dreams of creating a refuge where children will play and learn to dream. She dreams of creating opportunity for young adults to receive high school diplomas and job training so that they can extend the same, and greater, opportunities to their children. She dreams of providing working mothers with a safe and wonderful place for their children. Rev. Rosemary dreams of creating a community center and an affordable housing development in the truest meaning of the word *community*—shared love, respect, and hope. She boldly shared her dream with us on that first afternoon we met in Mississippi. After we toured DeLisle, she and Rev. Theodore asked if we would like to see the property where Mt. Zion was building a new church. I remember thinking, *Oh no, they had just started a new building project and now it has been destroyed.*

We only had time to drive by the site. It was a beautiful piece of open, slightly elevated land. A soft grass blanket covered it. It looked untouched by Katrina. A construction crew was putting up the frame. I asked Rev. Rosemary how much of the building had been completed before Katrina. "Only the foundation had been poured," she said and smiled. "We kept wondering why it kept raining so much" (as if to say that God had delayed progress before the storm).

"God gave me a vision for this land long before the storm," she continued. The church owned the twenty-eight acres we were looking at, and she wanted to develop twenty acres of it for housing and an education/community center.

"I want to provide adult vocational training during the school hours so young adults can get their GEDs (General Education Development, a high school diploma equivalent) and learn computer skills. And then in the afternoon the schoolchildren will come and use the facility. There is nothing like this in our community.

"I don't want something like the projects. I want a neighborhood. Some will not be able to afford to buy a home, so we want to make some rental apartments or something like that."

As I listened, my mind went back to the parable of the mustard seed that had given me hope in those first days as I watched the horrible images of Katrina's devastation from hundreds of miles away. I thought about how I had seen new possibility for the Children of the Storm in that parable. I said nothing, but I knew that Rev. Rosemary's vision and my hope were bound together.

A few days later Ellen, Shantrell, and I were driving to deliver some gift cards to a Gulfport neighborhood and I asked Shantrell, "What would you do if you could do anything and didn't have to worry about making money?"

"I would like to do something for the kids here. I want to teach them self-respect. I want to teach them how to dream." She told us how the "big-city" problems had migrated to her small community—that there never used to be teen pregnancy or drugs.

"I have seen that when it comes to kids, if you show them something else, they will *do* something else. I think one of the reasons there is so much teen pregnancy is because these girls don't have self-respect."

"What made the difference for you and your life?" I asked. "How did you end up being so driven and successful?"

"My mother made me believe I could do *anything*."

Shantrell wants to give her community what her mother, Carol Henderson, gave her—the confidence to dream.

Ellen was listening quietly. She makes her living by talking on the radio and she's very good at it. She always has something to say, but this time she didn't say anything. She was thinking.

The next day Ellen joined the conversation in her own way: She started making phone calls to see how we could help Rev. Rosemary realize her vision. She talked about the plan on her radio shows and posted the information on the Talk Radio News Service website. She called Bruce and Michael, her brothers in New York who had generously financed our relief efforts.

Rev. Williams had a vision, but now she had three other people (Shantrell, Ellen, and me), who were committed to that vision as well. An unbelievable chain of events and circumstances had put the four of us together to achieve something that was too big for any one of us even to consider attempting alone.

We had all the passion and commitment in the world, but no money. A wise man once told me: money is oxygen. We knew that we could not survive without it. Ellen and I began looking for money for the vision; Rev. Rosemary and Rev. Theodore continued to care for the immediate needs of their community; and Shantrell educated herself in the business of community development. She tracked down local officials, began the zoning paperwork, and searched for potential contractors—all this while she maintained her law practice, which, due to the hurricane, had unintentionally become a pro bono practice. She cared for her children, Mason and Taylor, and fielded constant calls for help from Katrina survivors at their wit's end.

I became increasingly frustrated as I sought funding. We knew that the development would likely be a patchwork quilt of various funding sources. We hoped we could get FEMA public assistance money for roads, sewer, water, and electricity because this housing would replace what Katrina had destroyed. We were told that these funds might be available at some point down the road, but the counties and cities had only been given FEMA funding for debris removal and security. We would have to wait.

I canvassed every federal agency that could possibly have something to do with Katrina recovery: FEMA, the Department of Housing and Urban Development, the Small Business Administration, the Department of Labor, and the Department of Agriculture.

To my surprise, the Department of Agriculture immediately responded to my questions. Ag personnel answered my calls and followed up with me to ensure that I had gotten the information we were looking for. After banging my head against the FEMA wall for more than a week without getting through to a human being, I was overwhelmed with gratitude for the Department of Agriculture.

The department quickly connected me with Florida Non-Profit Housing, Inc., one of its contracted rural housing agents that handle the southeastern United States. I spoke with Ms. Cheryl Wilkins and began to see some light at the end of the tunnel. "I love my job," was the first thing Ms. Wilkins said. She empowers nonprofits to help people in rural communities become homeowners. The new owners participate, working side by side with contractors to build their homes.

"Many of our homeowners are single mothers, but you would be surprised at how much they are capable of accomplishing— and *they* are surprised at what they can accomplish. They learn how to do repairs at the same time they are helping build their home."

Ms. Wilkins put me in touch with Ms. Daphne Wade, her representative in Jackson, Mississippi. A week later, back in Mississippi, Shantrell and I met Ms. Wade halfway between Gulfport and Jackson. She spent two hours walking us through all the details of the housing program and explaining how Mt. Zion could qualify to administer the program. Since then, Shantrell and Ms. Wade have worked together to see how this program can be used to help Harrison County.

The next morning Shantrell and I met with Rev. Rosemary and Rev. Theodore at the Waffle House (the only open breakfast joint in town) to map out exactly what they had in mind for the housing development and the community center.

To Shantrell's surprise, Rev. Theodore did most of the talking. She joked in the car afterward, "I've never seen him so talkative. He's normally the silent partner in that relationship." Rev. Rosemary's vision was contagious.

I asked whether they knew of any additional land around their twenty-eight acres that might be for sale. They smiled with a look that indicated they liked the sense of optimism implied in my question. Rev. Rosemary was quick to reply, "Well yes, there's that piece of land across the street, I believe. It's been for sale for a long time, but I don't know anything about it."

That weekend Shantrell, Myrick, and the kids went to see a Jackson State football game and I stayed behind. I drove over to the new church property after Sunday service at Mt. Zion. I wanted to see if I could find which piece of property was for sale. I enjoyed being on the church's land. It gave me a sense of peace and hope I had not felt anywhere else in the Gulf Coast.

As I was getting ready to leave, I noticed two men across the street on the porch of a small house. I saw an old sign that was so worn from the elements you could barely read the words FOR SALE BY OWNER. I crossed the street, introduced myself, and told them I was interested in buying the property in order to build a community center. I also briefly described what the church had in mind for the housing community across the street.

The younger man, Dr. Chuck, reminded me of a Marine. He was clean cut, physically fit, and extremely friendly. He told Mr. McKay, "You give this lady a good price on that land. She's with the church." As he was leaving, he asked Mr. McKay, who hauled gravel for a living, if he was doing all right and if he needed some extra work. Dr. Chuck took Mr. McKay's number and said he would call him.

As it turned out, Dr. Chuck was a potential buyer. From what I could gather, he had lost his medical practice in New Orleans due to Katrina, was originally from Harrison County, and planned to come home.

Mr. McKay showed me the house he had lived in before he moved in with his daughter and her family. He had gotten too old to live by himself. Then I learned his life story. He had come to this town with "nothin' but these two hands," he said as he opened his palms. He said he "hauled loads" and that he and his wife didn't go out to eat for seven years in order to save money and build his business. "She lives in Birmingham now, but I've got surgery on the twenty-sixth of October and she'll be right by my side. She's a good woman."

"Now," he said, "I need to tell you upfront that I'll need two checks when we close—one for me and one for her. I don't hide nothin' from her." We talked for over an hour. I told him that I intended to buy his property, but that I needed to do some research and I would get back to him. "You know," he said as I was leaving, "that fella Chuck was almost two hours late? I saw you across the street when I was waitin' for him and thought,

'Oh I hope she's comin' over here to buy my property,' and then I saw you get in your car when he pulled up and I said, 'Dang.'"

I ended up buying that property after a good deal of negotiating with Mr. McKay and his wife. Shantrell was handling the legalities and I was back in New York. As the purchase agreement was about to be signed, Mr. McKay did not want to guarantee in writing that the home was free of defects other than those we had previously discussed. For about ten minutes we went back and forth over the same arguments.

"Mr. McKay, you are telling me that the house is fine and that everything other than the water pump and a few parts of the roof are fine, so I'm just asking you to sign your name to back up your words. If everything is fine, why won't you sign your name?"

"But what if something goes wrong six months down the road? I can't afford to fix it."

"I don't expect you to. I'm saying that at the point of sale everything is good. You know—when I take possession, not six months down the road."

"OH! I get it. Well, I'm dumb you know. I told you I only had a sixth grade education. I'm not a big city wheel."

"Yeah, Mr. McKay, you're dumb like a fox."

He laughed and we had a deal.

After talking with Ms. Wilkins and Ms. Wade, we were encouraged by the possibility of help through the Department of Agriculture, but we knew their program had no new funding for Katrina and it would be a long-term plan. The needs, however, were immediate.

Ellen continued to raise private money for the project. She does not like to ask for money, but Ellen is obsessed with taking care of others. (In restaurants, she makes a deal with the waiter before the meal is served to ensure that she gets the check.) Ellen is always the one giving, not the one asking. Though valiant, Ellen's fundraising efforts had not yielded hard cash as quickly as she desired. It's difficult to convince people to give money to a cause they think is already receiving large sums of aid in the form of federal and private dollars through the Red Cross and other organizations. She was becoming increasingly frustrated and depressed.

"I think I have an eyelash in my right eye that has been driving me crazy for four days," Ellen told me when she arrived in New York one Sunday, back from a business trip. She attempted to wash the eyelash out with saline solution, but the irritation didn't go away. We got in the car and began to drive to Washington, D.C., as we had originally planned to do that afternoon.

Ellen's eye got worse and worse during the four-hour drive, and she decided that she needed to go to an emergency room. By then we were about forty minutes outside of Baltimore, Maryland, so I suggested we call a friend of mine from the Air Force Academy who is now a doctor at Johns Hopkins Hospital in Baltimore. Dr. Regina Brown arranged for us to go straight to the Wilmer Eye Institute at Johns Hopkins. It's the only emergency eye facility in the region. By the time we made it to the hospital, Ellen could not read the big "E" on the eye chart. She was blind in her right eye.

We soon learned that she had a partially detached retina. The doctor attempted to repair the tear using laser surgery. The repair failed a few weeks later, and I was already back in Mississippi when Ellen discovered that she needed major surgery. I flew to Baltimore, and the next morning I drove Ellen back to Johns Hopkins Hospital. She went to pre-op and I paced the halls.

I needed to pray. I thought of the "prayer request" part of the Mt. Zion service I had attended just three days earlier. As people voiced the names of those to remember in prayer, the congregation responded in unison, "The Lord is able." I decided to call Rev. Rosemary to ask her to pray for Ellen. This was a relatively desperate move for me. I had decided after Iraq that it was ridiculous to think that God stops everything to cater to my special needs and requests—or anyone's for that matter. My thinking represented the logical side of my brain, but my experiences in the Gulf Coast had awakened my faith, which transcended the limitations of my reasoning.

I dialed Rev. Rosemary. It was 7:30 AM central time. She answered immediately. I explained what had happened to Ellen and asked her if she would pray for her. "Let us pray right now," she said. "God, Ellen has a great task ahead of her. She will need her physical vision as well as her spiritual vision. I pray that you empower the hands of her surgeon so that he does your work, and fully restores her vision." She prayed for me as well and thanked God in advance and we said good-bye.

Ellen came through the surgery with flying colors, but her physical vision is yet to be restored. We remain hopeful. Ellen is convinced that she lost her eyesight because she was so disturbed by what she saw in Mississippi, coupled with her inability to make all the pain and suffering go away.

Our struggle to help the people of Mississippi has reminded me of something that happened to my brother Chip and me when I was four years old and he was six. Chip decided that we could make some money selling Kool-Aid. We found a cardboard box to use as our stand, and we wrote KOOL-AID 5 CENTS on it with a black Magic Marker. Our plastic pitcher was full to the brim with grape-flavored Kool-Aid. I can still smell it. I was very excited. My brother was letting me participate in one of his adventures, and it was an opportunity to make money (which I have always loved).

We had not even served our first customer when some older kids rode over on bicycles and started circling our stand and making fun of us. I was getting nervous. One of them suddenly kicked over our stand and the grape Kool-Aid spilled everywhere. I was outraged. I could not believe this was happening to us. It wasn't fair. My hopes were shattered, and the worst part of it was that we were helpless. They were so much bigger than we were, and there was no way we could fight back. I wanted to cry, but I saw the look of humiliation on my brother's face. He knew we were beat. He knew he couldn't defend me. So I held back the tears.

I now look back on that moment from a different perspective. Instead of humiliation and despair, I now find a lesson of hope and unity. The thing that made our short-lived Kool-Aid venture so special was my brother. Together, we were doing something to better our lives, and it didn't really matter that our first attempt failed miserably.

I view the setbacks and barriers encountered as we work together to realize Rev. Rosemary's vision for her community the same way. The people of Mississippi have encountered many bullies in their lives, particularly since Katrina struck. Their strength, however, is in their commitment to each other and to a vision—and that vision will outlive the memory of any one storm.

fifteen
MANAGING PERCEPTIONS

"Politics is managing perceptions. Governing is managing reality."

—*Chip Espinoza*

I had a few days of vacation remaining when Ellen and I returned from DeLisle. Ellen suggested that I spend the first week of October on Capitol Hill reporting on our trip to as many members of Congress and their staffs as possible.

"That's a total waste of time," I reacted to Ellen's suggestion. I often wonder if the whole enterprise of government has warped into one giant public relations machine. Our government feels more like a dramatization of government where creative geniuses spin the special effects of language and message to create a perception that they are governing. Meanwhile, the reality is that our government is spending and spinning.

The Marines I was embedded with were at the tip of the southern spear; the First Tank Battalion led the invasion into southern Iraq. I covered the war from the vantage of a member of the TOW (tube-launched, optically tracked, wire-guided) missile platoon. TOW missiles are tactically designed to "kill tanks," but they are mounted on small, flexible, and fast Humvees, so their mission is often flexible. As I was leaving the platoon to go back to the United States, "Gunny" Sgt. Thompson, a career Marine only a few months from retirement, said, "We're sitting ducks. Go tell 'em how messed up it is over here." I attempted to do just that my first few weeks back from the war.

It was almost immediately apparent to many Marines in Iraq that the strategy of the war was flawed and that there would be grave consequences for the U.S. military as they attempted to execute that flawed strategy.

The best way to sum up the war strategy was, cut off the head and the body dies. Remove Saddam Hussein and his command element and the rest of the regime would become good guys. I had been warned that the premise of the strategy was false six weeks before the war started.

I took my regularly scheduled check ride at the United Airlines Training Center in Denver the third week of January 2003 and then left for Israel where I met Ellen to help her cover the Israeli elections. I was able to take time off from United because they had just declared Chapter 11 bankruptcy and were drastically cutting the schedule.

Ellen went back to Washington, D.C., and I spent another week in Israel before traveling to Amman, Jordan. I was waiting for the war to begin. We were not sure if Talk Radio News Service was going to get a coveted embedded journalist slot. As a backup, I attempted to get an Iraqi visa and go as a non-embedded journalist. It was rumored that the embedded journalists had to be cleared from Kuwait, so I was also trying to obtain a Kuwaiti visa.

The Iraqi visa was a long shot, but I assumed that it would be easy to get into Kuwait. I was wrong. Each day I checked on the status of my application and each day I was told, "Perhaps tomorrow, *insha'Allah* (God willing)."

During the day, I walked the streets of Amman talking to businessmen, shop owners, workers, low-ranking government employees, and anyone who was willing to have a conversation with me. At night, I reported on my findings to the talk show hosts across America for Talk Radio News Service.

I found no shortage of Jordanian hospitality or insight. Men and women alike, many of them Iraqis living in Jordan, were happy to share their views on the future of their region—particularly given the inevitability of war in Iraq. They predicted with uncanny precision all the troubles the United States would encounter in Iraq, including predicting that we would not find weapons of mass destruction.

One man who worked at the local movie theater said, "Weapons of mass destruction? How you say . . . old car?"

"Yes," I said, "old car."

"Saddam military like old car that needs repair. He no have these type weapons."

I asked an Iraqi woman who was working in Jordan but who still had family in Iraq, "What will the government be like if Saddam is removed?" She said, "It will be a mess. There are too many different groups who want power."

Some of the best security analysis came from a businessman who specialized in regional travel. "Iraq is like a two-headed snake," he said. "You cut one head off and the other remains. The Baath members will still be there."

The most disturbing prediction I heard was from a man who was in charge of the Jordanian equivalent of a visitor's center. I came across this office on one of my daily walks. At first I thought it was closed; it looked deserted. I tried the door. To my surprise, it opened. I picked around some of the travel brochures. Everything was covered with a film of dust. The Intifada in neighboring Israel had exacted a heavy toll on Jordanian tourism.

I could hear men arguing in the next room. I had been in the region long enough to know that shouting did not necessarily mean there was a fight. Among Arabs, shouting is a normal way of communicating one's point. It wasn't personal, particularly when shouted over tea and cigarettes.

"*Marhaba*," I said (Arabic for hello). "Excuse me. I apologize for the interruption. I was wondering if I could have some of your time. I'm a journalist from the United States." The director of the office, a man named Tariq, invited me into the room and then took me to his adjacent office. He graciously entertained my questions for over an hour.

The last thing he said to me was, "I'm sorry, but I think this will be your second Vietnam. No, there will not be as many deaths, but you will not be able to determine friend from foe. They will not fight you in the open. They will hide with the people and they will fight you. This will be a very long fight."

Tariq was balanced; he believed that Israel had a right to exist (my personal litmus test for someone's sense of balance), and he was pro-democracy, but only genuine democracy, as in one person, one vote, not just a pretend democracy made up of a pro-United States head of state and a powerless parliament. Yet he was comparing the pending war in Iraq to Vietnam, a war that had inflicted wounds so deep they are still bleeding.

Most of what I know about the Vietnam War I learned at the Air Force Academy. Some of my instructors had flown in that war. They instilled in me an aversion to war—especially war against an enemy who could hide easily within their population.

Five weeks later I was sitting a few kilometers from the Iraqi border with two Marines, Corporal (Cpl.) Francisco Blea and Private First Class (Pfc.) Sean Paul, in a U.S. Marine Corps unarmored Humvee with a TOW missile mounted through the roof. We were having a philosophical discussion about the pending invasion. I commented that the way our nation would be viewed from here forward was in their hands and depended on their judgment—their determination of who to shoot and when to shoot.

When I'd said my piece, Pfc. Paul asked Cpl. Blea, "Should we tell her?"

"Tell her what?"

"Should we tell her what they said in the intel brief?"

"Sure. Go ahead."

"Those Iraqis you're worried about over there are really Saddam's guys," Pfc. Paul said. "They're taking over civilian homes and putting on civilian clothes." Tariq's words, "You will not be able to determine friend from foe," echoed like a jackhammer in my head. The Humvee got very quiet and our conversation ended. There was a silent consensus: in a word, we were screwed.

But this intelligence report did not translate into a change in tactics. Within the first hour of crossing the Iraqi border (the Breach, as the Marines call it), my platoon was embedded in the civilian population as though we were part of the neighborhood. The Marines were on edge knowing that at any minute the nice white pickup truck driving beside us or the Iraqi walking by us could be one of Saddam's forces, blending in perfectly with everyone else.

Civilian vehicles weaved in and out between the Marines' Humvees. "What is this?" Cpl. Blea asked, again expressing his displeasure at an Iraqi truck that cut in amid the small convoy. "This is too weird," he muttered.

"You've got to be kidding me. We're going to spend the night here in the middle of this neighborhood?" I said as we pulled into the area.

"Looks like it." Pfc. Paul was nineteen years old. He had graduated from high school the spring before the war, and he seemed to have internalized an often-stated military motto:

"Mine is not to question why, mine is just to do or die." Cpl. Blea, on the other hand, was one of the oldest and most experienced in the platoon—at the age of twenty-four. Back at the camp in Kuwait, while we waited for the order to invade, Cpl. Blea read a Shakespeare anthology. The best word I can use to describe Cpl. Blea is *wise*. His actions seemed to be based on a fund of knowledge that extended well beyond his years. Perhaps it was the Shakespeare.

Cpl. Blea had a sign over his sleeping bag that read, VICTIM OF STOP LOSS. (Stop loss is a "personnel retention tool" used by the Pentagon that allows the military to retain members beyond their commitment. Stop loss has been described as a backdoor draft.) He had been scheduled to separate from the Marine Corps and begin classes at New Mexico State University the previous fall. The government had other plans for what would have been his freshman year.

As the Marines traded watch duty back and forth that first night in Iraq, I slept. I was awakened by the choking smoke from the oil-field fires. The thick smoke and fallout from the fires made breathing difficult and left a blanket of tiny oil droplets on the Humvee.

Some Marines were bringing over about a dozen Iraqi prisoners. I went to take a look, and a Marine told me, "Sorry, ma'am, you can't talk to them." The prisoners were divided into three groups that I assumed were enlisted, noncommissioned officers, and officers. The enlisted looked underfed and had thin, scraggly mustaches. They were shivering from the cold and they looked frightened. The noncommissioned officers looked a little better fed—a little pudgy, even. Their mustaches were thicker. The officers were Saddam look-alikes; they sported his signature mustache, were virile-looking and had strong, square jaws. Their eyes were full of anger. They were not shivering; they sat, hands tied behind their backs, with a stiff-necked air of defiance.

Not wanting to make a freak show out of them by staring for too long, I walked back to the Humvee. As I stood by the Humvee, fumbling around in my backpack in the front seat, I heard a loud explosion immediately followed by more loud explosions. "Fuck!" Cpl. Blea and Pfc. Paul said in unison. I jumped into the front seat and they started to drive to find a position from which they could return the fire.

Having been a U-2 reconnaissance pilot, I wasn't used to driving into the fight. This required an instant and unwelcome paradigm shift in my thinking. We drove right by the Iraqi prisoners I had seen earlier. The enlisted and noncommissioned officers were ducking down as though that would save them from the impending doom. The officers, however, were still sitting as erect as ever—only this time with big smirks on their faces.

"I don't see these Iraqi officers handing over the keys to a new government in Iraq anytime soon," I thought to myself. The Marines were ill equipped to counter this attack launched from within an Iraqi neighborhood. They had TOW missiles and tanks—using either would have resulted in collateral damage for the entire neighborhood. As the Marines continued to draw fire, they requested permission to fire back in the general direction of the attack, but were repeatedly denied. The rule of engagement at the time was that the Marines could not fire unless they could positively identify who was firing at them.

Suddenly, another threat appeared. Three Iraqis in civilian clothes with rifles in their hands were running toward the platoon, as though they were attempting to get within firing range. One of the platoon members fired a TOW missile at them. It would not have been the textbook weapon of choice, but they didn't have anything else to use except an M16, and the Iraqis were too close for them to get out of the Humvee and into position to shoot.

The missile went erratic, as the Marines say. It became a giant $180,000 firecracker packing enough firepower to penetrate a tank, oscillating at warp speed just a few feet above the ground. The Iraqis were close enough for me to see the whites of their eyes as they got a look at that TOW missile heading in their direction. They dropped their weapons and ran faster than I've ever seen another human being run. They escaped both the TOW and the First Tank Battalion.

Sporadic shooting continued. Not being able to identify who was doing the shooting, the Marines grabbed their M16s and headed in the general direction of danger. "Cho, I think it may be best for you to stay here," Cpl. Blea told me.

Within seconds, our twenty-first-century shock-and-awe military had reverted to the centuries-old, tried-and-true, one man, one weapon, on foot army. Although shock and awe helped remove the head, it could not kill the body; those with too much to lose chose to fight on their terms. Gunny Sgt. Thompson knew that we did not have the numbers, the equipment, or the strategy to defeat an insurgency that began on the day of the invasion. And that is what he meant when he said, "We're sitting ducks. Go back and tell 'em how messed up it is over here."

I returned to America with those words etched in my mind. I had a mission. I wanted to report what I had seen. Ellen used her connections and relationships to secure meetings for me with members of the House of Representatives and of the Senate (both Republicans and Democrats), as well as anyone else in a position of influence who was willing to give a reporter some time.

I explained to them there was a dangerous organized element that was left in place while the Marines were ordered to move forward, and I thought that that element was going to cause problems for Iraq—as well as the military left there to stabilize it. I explained that the Marines did not have the proper weapons to check the resistance they encountered, and why a TOW missile was no match for an AK-47.

I told them that the Humvees had no protection. I told them that the Marines had been using sand bags to protect themselves from blasts. I told them about the equipment that was breaking down in the harsh conditions of sand, wind, and rain.

Everyone listened attentively with a contrived look of concern that must have been learned in a required course for politicians, Apparent Sincerity 101. They all nodded their heads, patted my back, and sent me on my way as if they were sending a child back to bed after a nightmare. My time had been wasted.

Tariq's admonition still rings true. The insurgency has become increasingly sophisticated. Terrorist immigration and recruitment make matters worse, particularly for the Iraqi people who are their primary victims. Still, there are no alternative solutions or plans emerging from Washington, D.C.

We have investigations and hearings and compulsive deconstructing and reconstructing of the evidence that led up to war, but none of that accomplishes anything for the people on the ground in Iraq. The complaints remind me of someone who goes into a business venture without performing proper due diligence and gets his comeuppance. Except that in the business world, when the venture turns out to lose money, the flow of money ceases until the business plan and/or the management team changes. With the Iraq War, money continues to be pumped into the "business" at a higher and higher rate.

According to a Congressional Research Service (CRS) report dated October 7, 2005, titled *The Cost of Iraq, Afghanistan and Enhanced Base Security Since 9/11,* "DOD's (Department of Defense) current monthly average spending rate is about $6 billion for Iraq based on the first nine months of fiscal year 2005. Compared to fiscal year 2004, those averages are 19 percent higher for Iraq."

Not only does money continue to be invested at a higher rate, but no one seems to know where the money is going. The CRS report goes on to say, "The DOD has not provided an overall reckoning of these funds by mission or military operation."

Our system of government, with power divided between the executive, legislative, and judicial branches, is supposed to function with checks and balances. Yet it seems to have morphed into a system in which Congress writes the checks and there is no balance in the checkbook (or elsewhere).

Katrina was a hauntingly similar drama shot at a different location. The Gulf Coast looks like a war zone. Billions of dollars are being spent, but no one knows where the money is going. The people are doing the best they can to get on with their lives, but it's difficult to do that without resources, and the resources are tied up in a bureaucratic knot somewhere. No one seems to know who is in charge.

Within a few days of Katrina's landfall, $62.3 billion were allocated for relief and recovery. Congress has appropriated and President Bush has signed Public Law 109-61 for $10.5 billion and Public Law 109-62 for $51.8 billion. The fund is now up to $85 billion, according to President Bush. "I want to remind people in that part of the world, $85 billion is a lot," the president told critics at his January 29, 2006, press conference.

Yet despite the president's assurances, over six months after

Katrina struck the Gulf Coast, neither public nor private money had yet reached the overwhelming majority of Katrina's survivors. Initially, some had received FEMA's $2,000 and/or a $2,300 rental voucher, but the funds to rebuild were somewhere between bureaucratic gridlock and commercial insurance purgatory. The federal government was waiting to see what the insurance companies would pay, but many of the insurance companies were taking their time and/or denying claims. I did not meet any businessmen or homeowners who had been approved for a loan under the Small Business Administration disaster loan program. Cities and counties were broke, scrambling to fill out debris removal receipts.

FEMA trailers are an excellent example of government waste and mismanagement. The government is spending hundreds of millions for temporary measures when they could empower local communities to rebuild permanently instead. *The Washington Post* reported on November 25, 2005, that the federal government plans to spend close to $2 billion on temporary trailers.

The week prior to Thanksgiving, Ellen and Shantrell paid a visit to a FEMA trailer lot to see if they could find any information on how the trailers are distributed, because there was no discernable method to the process. They were trying to help a disabled schoolteacher, who was still living in a tent, obtain a trailer. One young man at the lot was very friendly and talkative and told them that the Bechtel Corporation operated the center. He told them that dozens of sets of keys were missing and that was part of the reason that the lot was full of new, uninhabited trailers. Suddenly a security guard showed up to chase them off the Bechtel property. Ellen asked him, "Can we talk to your supervisor?" He said, "George Bush is my supervisor. If you have a question, call him."

The security guard pointed Ellen and Shantrell in the right direction; the president is ultimately responsible for FEMA. The way relief and recovery is supposed to work is that Congress appropriates the money and the executive branch gets the work done.

FEMA, under the Department of Homeland Security, is the primary federal agency responsible for disaster assistance. Volumes have been (and will continue to be) written about what FEMA did or did not do in the immediate aftermath of Katrina; the fact remains that there is still an administrative crisis in the Gulf Coast area caused by a lack of leadership in the executive branch of government.

FEMA's incessant lowering of expectations, bureaucratic tail-chasing, and general neglect can all be summed up in one useful phrase I learned from Patrick Markee of the Coalition for the Homeless in New York City: "administrative disentitlement." It is systematic slow rolling.

While "entitlement" has become a bad word in our society (to say someone has an "entitlement mentality" implies that they want to sit back and watch others do the work—they want a handout, and they *expect* a handout), I did not meet one person in Mississippi who felt "entitled" to government assistance. They simply knew that survival was questionable without it.

One morning I was running through Shantrell's neighborhood. Most of the debris had been cleared, and the families had manicured what was left of their small yards so you could focus on the homes instead of the trash. I noticed that it looked a lot like the neighborhood I lived in as a child in Hobbs, New Mexico. The homes were modest, probably three bedrooms at most. I ran by an elementary school and could see the children at their desks through the window. I pictured myself sitting in that classroom.

I wondered what would have happened to my family if something like Katrina had come through Hobbs when I was the age of these schoolchildren. My mother and stepfather both worked. We rarely ate out unless we were celebrating a special occasion, such as someone getting straight A's (although that was not me). My mother fed a family of six on a pound of hamburger or ground deer meat nearly every night. We drove to all of our vacation destinations, which were always at a campground or a relative's home.

Our social life consisted of church and the sports my stepfather coached, and my brothers played at the Boys Club. My parents paid their taxes and their tithes. We thought of ourselves as middle class. Overnight it all could have collapsed. We, just like the families of the Gulf Coast, would not have had the cash or credit to rebuild on our own. We would have needed our government to govern.

Instead, billions of taxpayers' dollars are invested in "known contractors," like Bechtel, under the veil of expedience while local businesses and individuals in the Gulf Coast cannot even borrow the money to recover.

While many in our government are focused on America's battle of "good versus evil" in the "Global War on Terror," they have abdicated our moral responsibility to our fellow citizens at home. Self-reliance, individual responsibility, and the strength of the private sector are hypocritical euphemisms in light of the moral crisis brought on by waste and fraud, corporate welfare and greed, bureaucratic incompetence and neglect.

sixteen

PRESIDENTIAL VISIT

"Rhetoric does not drive away the cold."
—Mississippi Gulf Coast high school teacher

"MOVE! MOVE NOW! LADY, GET THAT CAR OUT OF HERE! NOW! I SAID NOW!!! YOU ARE BLOCKING THE MOTOR-CADE!" A police officer was blasting these orders at us over his bull horn.

Our timing could not have been worse. We were initially directed forward into the neighborhood of Gulfport, Mississippi, where former President Bill Clinton was going to speak, but suddenly the officer changed his mind. He must have heard over the radio that the motorcade was arriving. He said, "STOP!" but we were already committed and could not turn around because the motorcade was in the way.

Motorcades create their own special "no-drive zone," similar to the Iraqi no-fly zones I used to monitor as a U-2 pilot. It seems no one is allowed to move when a presidential, or, in this case, former presidential motorcade is in motion nearby.

Shantrell was unflappable as the officer blasted her through his speaker against the car window so loudly that my ears hurt. She casually looked around for an escape route while the motorcade waited. We parked on the side of the road and walked the rest of the way.

People from the neighborhood had already lined up along the side of the road as if there was an invisible rope preventing them from crowding the former president. He started at the far end of one block of the neighborhood and made his way down the block. It seemed as though he was taking a casual stroll through the predominantly African American neighborhood. He took his time and physically touched nearly every resident. People waited patiently.

Many of the people knew or recognized each other from town. The conversation gravitated toward the same topics—Katrina and family.

"How did you make out?"

"How's your mom?"

"How are your babies?" In the South, there are "babies" and there are "*gran*'babies." From what I informally observed, these terms are used to describe children until they reach the first grade. Then they become "kids," as in, "How are your kids?" or "How are your *gran*'kids?"

As we waited in line, I felt as though I was part of a large family reunion. Everyone was hugging each other, laughing, and swooning over each other's babies, gran'babies, and kids. They spoke about the damage to their homes only briefly.

The conversation usually went something like this:

"How you doin'?"

"Oh, we're doin' fine. We had four feet of water and a tree in the roof, but we're doin' fine. How are you doin'?"

"Doin' fine" is almost an expression of faith. It seemed as though to say otherwise would be to admit defeat. I had visited the uninhabitable homes of people who said they were doin' fine. I could see that they were doin' anything but fine. They were patiently waiting on their insurance and/or FEMA to come through with enough money to start rebuilding. They had applied for a loan through the Small Business Administration, but again, they were waiting to hear back.

Some felt guilty that they had fared better than others had. One person told an elderly couple who had lost their home, "I'm so sorry. I feel so guilty." The husband and wife said in unison, "Don't!" They thanked the man for his concern and said, "Bless you for saying that, but it's really not necessary. We're just happy that some people made out okay."

Shantrell and I were standing next to Mr. Purvis McBride, the director of the Gulfport Boys and Girls Club. He and his volunteers were holding up pictures for Bill Clinton to sign. The photos were of youths who had benefited from the Boys and Girls Club and now were accomplished adults.

Parents and grandparents asked the director, "When's the center going to reopen?"

"We're sure tryin'. I hope real soon."

"Here's my number. You call me when it's ready to paint," one older woman said. She laughed. "My boy is driving me crazy. He needs that center opened as soon as possible."

Many were holding copies of *My Life*, Bill Clinton's autobiography, for him to sign. Old and young were holding the books in the air proudly. Some of the copies were worn and well read.

Shantrell and I were at the end of the line. President Clinton shook Shantrell's hand and then made his way over to a small open picnic tent where some of the elderly and disabled were waiting for him.

I followed along to position myself with the rest of the press, but was soon stopped by a member of the Secret Service. One of the local community hosts interceded and said, "She's allowed," and gestured for me to join the press corps.

Governor Haley Barbour, his wife Marsha, and Gulfport Mayor Brent Warr sat next to the former president. The neighborhood's residents were sitting in a circle flanked by another row of local people who were standing.

The governor thanked the community and said, "I 'preciate all the hugs," and then he turned it over to Bill Clinton.

"I just want to say two things and then open up the conversation. When I was a boy growing up in Arkansas, we took one out-of-state vacation my whole life and it was to New Orleans, Gulfport, and Biloxi. So I saw a lot of what's been taken down. Former President Bush and I have raised $90 to $100 million to distribute across these three states."

He explained that they did not want to duplicate the efforts of others and wanted the money to help the largest number of "all of you." He said that he would take information back to Washington and that Congress was due to pass an aid package in the next four to six weeks. He said the president wanted to know what, if anything, the federal government could do differently or better. He said that it wasn't about politics or Democrats or Republicans—everyone just wanted to do the right thing to help our fellow human beings pull their lives back together. He thanked everyone for the hugs he had received and said he was in a "better humor than I was when I got out of that airplane." He then opened the meeting for questions.

"How was the last doctor's report?" was the first question. It struck me that here was an opportunity for these people to vent for the first time, and they were more concerned with his health than with their own welfare. He said that his doctor told him he was in the top five percentile of health for men his age.

One woman who had flood insurance spoke. Six weeks after calling her insurance company, she still had not seen or heard from them. She had bought flood insurance even though no one thought she needed it—she had been there for forty years and there had never been a flood. "This is where we are," she said. "We are stuck."

Governor Barbour agreed. "We're gonna need federal help.

It's hard to believe we're a couple of miles from the flood zone," he said, gesturing toward the neighborhood that Katrina had soaked under four feet of water.

As people continued to tell similar stories, Clinton listened. He said he felt partially responsible for the fact that they did not have flood insurance, because the federal government decides who is in a flood zone and who isn't. He said an expert decided this and that "we underestimated the power of the storm."

"Did you get any money from the federal government? Two thousand dollars or something like that?"

Some people nodded and one woman said, "No. No. I need to speak. I have lived here all my life. The people you see raised us. I am a teacher. I *may* be able to rebuild, but they will never get back to a basic living without some type of assistance. People are hurting."

Clinton asked, "Can you tell me what reasons were given (for not getting the FEMA money)?"

"If you were not displaced at the time you made the call to FEMA, they played a word game," she said. "I'm telling you what I know. They asked, 'Are you living comfortably in your home?' Well, I have a ceiling falling in, but relative to others I'm comfortable."

President Clinton asked if there was someone who had federal authority to say yes or no concerning the money. No one had an answer.

Brent Warr, the mayor of Gulfport, brainstormed out loud that perhaps he could try to get a cruise ship pulled into the area for temporary shelter. He was told previously that the ship had already been contracted by FEMA and only one hundred people were living on it, less than 10 percent of its capacity.

A woman who appeared to be about 55 years old spoke out. "I don't need a ship. I need some sheetrock." She raised her arms and pantomimed rebuilding her own home. "I just need the materials. I can paint. I can work. I'll get the work done." The crowd cheered wildly in agreement. President Clinton smiled as if he was proud of their self-reliance and resilience.

A young minister spoke of the immediate health hazards from mold. He said he had gone house to house in the neighborhood of two hundred homes to tell people about the Clinton visit and only counted fourteen FEMA trailers. He then said he was sick by the end of the day from the mold. "This is inhumane."

President Clinton said, "I want you to know that I don't mind that you are angry and upset. I'm not in office but if I were, you would still be having some of these problems. There's so much damage over so much square footage in such a short period of time."

He began to brainstorm with the group to see if they could come up with some solutions. The consensus was that people wanted the materials to rebuild their own homes and they would get the work done. "We don't need to wait for trailers," one woman said.

President Clinton speculated that they might be able to get companies to advance materials with a guarantee that they would be reimbursed.

Governor Barbour jumped in and said, "The fact is that there are seventy thousand homes destroyed in Mississippi. We are going to be building back homes for years. It's gonna take time. There's gonna have to be temporary homes." (Some estimate that that number has grown to over 103,000 due to the mold problem.)

President Clinton redirected the conversation to solutions. "Your number one priority is getting back into your homes as quickly as possible. If and when Congress passes this (aid package), one thing you've convinced me to do is to call Hillary when I get home tonight and tell her to do whatever she can to not take six weeks to pass this bill. They could pass it in six days if they took a notion and released the money." He then discussed the possibility of getting with former President Bush to use the Bush-Clinton Katrina Fund to provide matching money to the state.

The neighborhood was buzzing after the visit. One woman said to Shantrell, "President Clinton touched this arm. Be careful now, I'm not going to take a shower today."

President Clinton is considered, even by his most hardened critics, to be a masterful politician. I witnessed a masterful human being. He did not blame anyone or take advantage of the situation—he went the opposite direction by assuring that their troubles were not the result of intentional neglect. He did not make promises. Instead, he sat back in his chair, as though there was no other place he would rather be or needed to be, and listened.

When the complaints became more intense, he interjected a question in order to better understand the concerns and reassure them that indeed he was listening. He did not piggyback on their anger with derogatory comments toward the effort or lack of effort of others. He assured the members that he understood their needs and was going to do what he could to voice those needs to officials who had the authority to address them.

As Bill Clinton spoke and listened, I began to understand what others call his "gift." He has the ability to inject calmness into a painful and volatile situation. President Clinton had been what my brother Chip would characterize as a "non-anxious presence" in an anxiety-laden situation.

As we walked back to the car with several others, there was a sense of hope that had not existed before the visit. Bill

Clinton no longer held the most powerful office in the nation, but that did not matter to this neighborhood. The fact that he had come all the way to Gulfport, Mississippi, to spend over an hour in a neighborhood of flooded homes with people who had no place else to go, made them forget about their pain—if only for a few hours.

A few days after Clinton's visit, Ellen forwarded an e-mail that said another president was coming to Mississippi. George W. Bush was scheduled to come to the DeLisle Elementary School. Ellen's news organization, Talk Radio News Service, is part of the White House Press Pool. I am not a designated White House reporter, so Ellen had to ask for specific approval for me to be part of the press corps pool covering the visit. Her request was denied, but that didn't stop me from trying another angle.

DeLisle Elementary is within walking distance of Rev. Rosemary's church. The night before the president's visit, I had been at Mt. Zion Church helping community members fill out SBA loan applications. Rev. Rosemary and Ms. Bowser said they would meet me at the school the next morning to see if by chance we might get to see the president.

"I called the school and they said to be lined up by 7 AM for security and road closures," Rev. Rosemary told us. The president was due to arrive sometime after 9 AM.

It was the second day of school for the children of DeLisle and Pass Christian. The Pass Christian schools had been destroyed. The middle schools and high schools from both communities would be using temporary classrooms set up in trailers on the DeLisle Elementary property, but were not scheduled to begin school until the following week. I was told that before Katrina there were two thousand students in kindergarten through twelfth grade among the combined schools. No one was sure how many kids would be back for the year, but it was assumed there would be at least one thousand.

There was a buzz in the air as I walked to the school after parking alongside many of the parents about a quarter mile away. The children had been told on the previous day that they might not get to meet the president, but there would be other visitors including Governor Barbour, senators, and congressmen.

The students, parents, teachers, school staff, and hopeful presidential greeters were made up of blacks, whites, Latinos, Asians, Republicans, and Democrats. One tall, attractive white minister wearing a "Texas" baseball cap said, "I sure didn't vote for him, but he *is* our president." Ms. Bowser and Rev. Rosemary were already there when I arrived. They knew everyone. Ms. Bowser had taught the fifth grade for over thirty years. Her old students hugged her. Some were mothers of the children now in school.

Ms. Bowser talked and joked with her former students. Angela Rice was one of them. She was there with one of her daughters, Lacy, who had also been one of Ms. Bowser's students. The three of them laughed and eagerly awaited the president's arrival. Angela's other daughter, Kaylee, a third grader, was waiting for the president inside the school gym.

Angela told us that Kaylee had agonized the night before because she was afraid her dress was not pretty enough for the president. Most of these children had lost all of their clothes in the storm. Many of them no longer had homes—they lived in campers and tents.

Angela and her husband had registered with FEMA three times because they kept getting dropped from the system. Her husband had lost his job due to the storm but found a new one, though it paid less and had no medical benefits. Angela normally worked also, but she had not been able to recently. She is expecting a baby in January and was bedridden due to bleeding behind the placenta. She insists that she is fine now. That day she and her teenage daughter stood outside waiting for the president for three and a half hours. She worried that Kaylee was not going to get to see the president. Someone told her later that he never made it into the gym where many of the children waited.

Ms. Bowser's neighbors, Billie and Jessica Hillier, walked over to give her a hug, and then one of them said to her with a chuckle, "I found some of your things in my yard. I'm saving them for you." Her expression then changed to sadness. "Do you remember our fluffy black and white cat? I can't find her." Another woman jumped on the opportunity and said, "Do you want another one?"

The excited children filed into the school. One boy wandered toward us and away from the metal detectors and began to throw up. "Poor thing," Ms. Bowser said. "He's just so excited to meet the president."

The president's motorcade arrived some time after 9 AM. We could not get anywhere near where he entered the school so we waited outside for him. The crowd of about seventy people tried to ask the Secret Service if they thought the president would come out and say hello. The officers repeatedly snapped, "He does what he wants." Some people asked if someone would make sure that he knew they were waiting. The answer they received was a short, "Can't do that."

One woman, Ms. Ruby Wallace, had a letter to give to the president that proved she was related to his mother, Barbara Bush. She asked, "Sir, will you give a letter to the president?"

"Absolutely not," the Secret Service officer told her.

"But I'm related to his mother," she implored, "and that means I'm related to him. I have documented proof. I'm trying to get it to him so he'll know it." I got the impression that she just wanted the president to know that he had family in the region. It seemed to be her way of telling the president, "If you ever need a place to stay down here, come on by." She even looked like Barbara Bush.

The crowd speculated that the president had to pass by us because the rest of the students were waiting for him in the gymnasium, and the only way to get there was by the pathway where we were standing. The Secret Service had directed us away from the other possible passage point, which made the crowd suspect that we had been duped.

The conversations were a variation of the same theme I had heard while waiting for President Clinton. "How you doin'?" "How'd you come out?" "Oh, I'm doin' fine."

Ms. Wallace, Barbara Bush's cousin, told me that her home had been totally washed away and had not been found. "I had eighteen to twenty-three feet of water over my home. I built it like a boat and no one knows where the boat floated." She was living in a camper. She told the story of her Chihuahuas surviving the storm. "The mama Chihuahua swam her two babies away when the storm washed away the house, and then she led them back to where the house had been. My neighbor found her and fed her water with an eye dropper and saved her, and then she saved her babies."

As time went on, the small crowd began to question the level of security and the fact that the president's motorcade whizzed by "doin' ninety to nothing."

Rev. Rosemary said, as though she felt sorry for the president for being so isolated, "That's no way to live. If I were the president, they'd just have to shoot me because I would have to be with the people."

Another woman had traveled more than an hour to see the president. Her elderly mother was in the car waiting. "They have to be like that, y'all. Everything's got to be secret," she said. "My momma's in the car cussing me out because she couldn't walk up here to see him."

"I was right over there in my second grade classroom when they told us JFK had been shot," a woman named Shirley Gates spoke out. "We put our heads on our desks and prayed."

"That was the saddest day ever in America. The whole world mourned, not just Americans," Ms. Wallace said.

"Why do you suppose JFK had so much of an impact?" I asked.

"He was charismatic. He really cared. And the whole world *knew* it."

Finally, after the governor and local politicians filed out and the Secret Service began to relax and take off their jackets, the crowd moved out to the road as a last-ditch effort to see if perhaps they could get a glimpse of the president in the motorcade. Again, he was sped away. One woman said excitedly, "I think I might have seen him."

I took a picture of two women, Ms. Nora B. and Eleanor Jones, who had made a huge sign that was bigger than they were. THANK YOU AMERCIA, it said, and it had a smiley face on it. As I filmed them they said, "We made it for you, America! We just want the country to know we are so thankful."

On the hike back to my car, I saw an elderly woman named Gloria Harshbarger who had also waited for over three hours. She had been holding a large American flag all morning. I asked her if she was okay. Earlier she was smiling, joking, and eagerly anticipating the president, but now she looked like she was in pain. "I'll be fine," she said with resignation in her voice. "I have two artificial knees. I live a mile away, but I called someone to meet me on that corner. Thank you anyway."

I was angry and disappointed that these people who had

been through so much had not gotten to see the president. I reflected on how isolated and hyper-managed the president must be not to have the awareness to take time at least to say hello to more people.

As it turned out, he only visited a few children who were located near his entry point. He never made it to the gymnasium. No other dignitaries visited the other classrooms. The teachers and children were disappointed, along with the crowd outside.

This was not a typical political photo-op rent-a-crowd. Most of these people had lost everything. All ages, races, and political affiliations had come out to see their president because they had such respect and reverence for him and the office he holds. It seemed like a waste that the taxpayers spent millions of dollars so that the president and his entourage of dignitaries could make an appearance for a couple of hand-selected classes at DeLisle Elementary. The White House Press Corps took their pictures and recorded his two-minute "speech," but most of the children, teachers, parents, and community—the people who needed him to lift their spirits—waited in vain.

According to the White House Press Pool report and White House website, the president met with local, state, and federal officials in a private meeting and then greeted a kindergarten class from 10:54 AM to 10:56 AM. They were on the road again within twenty minutes.

Perhaps if the president had taken more time to connect with the people of this community he would have felt compelled to do more to ensure that federal aid is not wasted, and that it ends up in the hands of those who will put it to the best use.

"Listening," my brother Chip counsels business leaders, "is responding." In the five months since President Bush visited Mississippi, the lives of the residents of DeLisle and Pass

Christian have not changed. And while former President Clinton may have called Senator Clinton the evening of his visit as he said he would, to encourage her to help pass legislation that would get desperately needed supplies to the Gulf, the relief had not arrived as of five months after the storm. The Thomas.gov legislative internet search engine, which tracks legislation under the Library of Congress, came up with 337 references to Katrina in proposed bills. Nevertheless, these intentions did not translate into relief for those in need.

Meanwhile, the Bush-Clinton Katrina Fund had not distributed any money three months after the storm. I was told by a Fund staffer a week after Clinton's visit that they were still wrestling with the IRS to get 501(c)3 tax-exempt status. The staffer warned me that no money could be given out until the fund had the proper tax status. Two former presidents, one of them the father of the current president, could not get their nonprofit approved by the IRS in three months!

As I drove away from DeLisle the day of President Bush's visit, I wished that the extraordinary people I had met that morning had been given the chance to meet (or at least see) their president. More important, I wished that President Bush could have met them. I believe that if he had spent time with them, things might be different. He would make them a priority as a matter of his and our nation's conscience. He would find a way to lead them out of this crisis.

How different this visit had been compared to President Bush's visit to Ground Zero after 9/11. He spoke with conviction, and he walked the grounds of the fallen and shared himself with the relief crews. While 9/11 and Katrina are different, both generated an enormous human crisis. Katrina's path of destruction is not confined to one part of one city—it spans ninety thousand square miles. Katrina's wake of suffering worsens daily as it becomes more and more apparent that no relief is coming.

I have seen the people of New York and the Gulf Coast rise above the suffering and give to one another. I have seen volunteers from all over the nation come to the rescue of both communities. I wish the government that serves these Americans had even a fraction of the commitment to them as they have to their government.

A high school teacher who lives in a tent outside her mold-infested home summed up the need for leadership: "The failure of the government in the face of this disaster borders on domestic terrorism for those of us in the face of the oncoming winter. Will the deaths from exposure be linked to Hurricane Katrina? Rhetoric does not drive away the cold."

seventeen

LOSING FAITH

I'm almost embarrassed to say that I had been up most of the night in anticipation of President Bush's visit. I prayed that his visit would inject leadership into this human crisis. Instead, the visit appeared, as several of the people who waited outside and inside the school remarked, to be just another photo op.

As I drove away I suddenly began to cry, and then I sobbed. I had to pull over. I'm normally controlled by reason, but I became paralyzed by an overwhelming wave of grief. I felt as though the tears were going to rupture my skull if I didn't let them out. As I wept on the side of the road I revisited the despair I felt as we invaded Iraq. While my government did not order Katrina's wrath, it was not going to fix it either. Hopelessness, anger, and bitterness flooded in. I feared that nothing would improve for Katrina's survivors. A spiritual "fog of war" started to form again in patches of disillusionment in my soul.

The war gave me an appreciation for the saying "There are no atheists in foxholes," but my experiences during and after being in a foxhole, albeit a twenty-first-century foxhole, shattered the foundation of my belief system; the lens through which I had viewed my world, God, and Country.

God played only a minor role in my consciousness while I was in Iraq. I heard God's name mentioned just once. As the platoon crossed the border into Iraq, Sgt. Elias Franco, a Marine reservist from my home state of New Mexico, uttered one short prayer over the radio: "May God help us." His words were spoken softly, but the tension in his voice carried over the airwaves straight to my soul.

Sgt. Franco had an eight-year-old son and his fiancé, Carmen, waiting for him back in Albuquerque. Like many reservists, he had left his own business behind when his reserve unit out of Amarillo, Texas, was activated. He had resigned himself to the fact that his locksmith business was unlikely to survive his absence. Sgt. Franco and I talked frequently during the ten days we spent in Kuwait, camped within about thirty-five kilometers of Iraq, awaiting the war orders. I respected him instantly. He had a reserved and humble manner that I have come to appreciate as the mark of a strong leader. He was quick to be of help and quick to calm the nerves of those who were getting restless from waiting in the Kuwaiti desert.

I noticed one day that he had printed "Psalm 91" on the thin elastic band holding the camouflage cover on his Kevlar helmet. My mother had asked me to read Psalm 91 during my last conversation with her before leaving for the war. It reads in part:

> You who live in the shelter of the Most High, who abide in the shadow of the Almighty, will say to the Lord, "My refuge and my fortress; my God, in whom I trust." For He will deliver you from the snare of the fowler and from the deadly pestilence; He will cover you with his pinions, and under His wings you will find refuge; His faithfulness is a shield and a buckler. You will not fear the terror of the night, or the arrow that flies by day, or the pestilence that stalks in darkness, or the destruction that wastes at noonday. . . .

Sgt. Franco's TOW missile platoon was invading Iraq a week later. It was still dark, just before dawn on the morning of the invasion. The battalion's combat engineers had blown a path through the Iraqi defensive berm and marked the invasion path with fluorescent light sticks. Just prior to crossing over what was left of the berm, Sgt. Franco keyed his radio mic and said, "May God help us."

Nothing was said in response as the platoon drove through patches of fog nestled along its path. The radio transmitter in the Humvee I was riding in crackled and went dead. It had been troublesome before, but there were no replacements available. The two Marines I was with spent most of the first day of the war without being able to transmit a call for help or a warning of danger.

The darkness of the predawn invasion hour was interrupted by the glow of burning Iraqi military machinery. Some of the operators of that machinery were still hanging—frozen in their last position as they had tried to escape.

There was a strange odor in the air, similar to that which lingered over New York City in the days following 9/11. *This must be what hell is like,* I thought to myself. I felt dread—there was no turning back. I knew that going to war in Iraq was no longer a subject of conjecture or debate; it was now a part of the history of my country. My instincts told me we had just opened the gates of hell. Then I told myself that I was just reacting to Iraq as though it was 9/11, and that this was different and that I would soon witness the liberation of a nation. I wanted to believe that everything was going to go as our government had told us it would. I tried to think like a child, confident that the adults in charge knew what they were doing.

God didn't come to my mind again until we were stuck in a sand storm four days later. We had parked next to an Iraqi neighborhood that was covered with fine sand that looked and felt like baby powder. The visibility was so bad that I could not see my hand at arm's length. The driver's side door had fallen off our Humvee the day of the invasion, and there was a giant open hole for the TOW missile and gunner. We had no protection from the elements. The Marines were so exhausted that the suffocating sand didn't seem to faze them.

I wrapped my lucky handkerchief around my face and prayed for peace—for an end to the war. It was my first quiet time since the invasion had begun. All I could hear was the howling wind, which eventually lulled me to sleep.

We stayed at that location for several hours. Finally, the blowing sand turned to blowing rain and we were a battalion of mud pies. The Marines drove through the night. The next day we had our first contact with the outside world when a car full of French journalists drove by our platoon. I told them to be careful because there were mines all over the place. I asked them if they had heard anything about a surrender. They looked at me with pity in their eyes that signaled they instinctively knew that there would never be a surrender. One of them finally said, "No. I'm afraid not."

I returned to the United States to find that there was very little, if any, possibility for the peace I had prayed for. The Iraqi combatants I had encountered and described in my radio reports continued to gain resolve to fight.

Our foreign policy was war, as officially stated in the Bush administration's preemptive strike doctrine. I returned to a pro-war country. Between a government that defined its leadership by the ability to wage war and a population that still wanted to avenge the attack of 9/11, prayers for peace seemed hopeless. Worse, some of the most vocal advocates of preemptive war claimed to be true devotees of the teachings of Christ.

I could not understand this. I knew that Christ's teachings had a different view of enemies, retaliation, and war than the people who were making, selling, and executing U.S. policy. Christ gave the world an alternative to violence:

> You have heard that it was said, "An eye for an eye and a tooth for a tooth." But I say to you, Do not resist an evildoer. But if anyone strikes you on the right cheek, turn the other also; and if anyone wants to sue you and take your coat, give your cloak as well; and if anyone forces you to go one mile, go also the second mile. Give to everyone who begs from you, and do not refuse anyone who wants to borrow from you.
>
> You have heard that it was said, you shall love your neighbor and hate your enemy. But I say to you, Love your enemies and pray for those who persecute you, so that you may be children of your Father in heaven; for He makes His sun to rise on the evil and on the good, and sends rain on the righteous and on the unrighteous. For if you love those who love you, what reward do you have? Do not even the tax collectors do the same? Matthew 5:38–46

The depravity of terrorism cannot be underestimated or minimized. I have seen its carnage first hand across the globe and in my own city. Yet our violent response to terrorism has introduced a magnetic field to our moral compass. We risk losing our nation's true north.

It has been said that history does not exactly repeat itself, but that it rhymes. Our modern Holy War—in which both sides believe that the ends, mutually defined as elimination of evil, justify any means—reminds me of the Spanish, Portuguese, and Mexican Inquisitions that my ancestors experienced. Professor Martin Cohen has translated the arguments used to justify the continued oppression of the "New Christians" (converted Jews) in his book *The Canonization of a Myth.* The records show "repeated assertions that the Jewish condition is irremediable and incurable because the Jews are inherently unrepentant and incorrigible." The church writers argued that "in a just war one may capture the innocent along with the guilty." The newly converted Jews were *less than,* incorrigible, not worthy of the dignity of the rule of law or the protection of innocents.

As I watched this twenty-first-century Holy War played out in Iraq, I became cynical and skeptical of anything that included the word "God." Reverend Stephen Bauman of Christ Church in New York City once said, "Some have reduced God to a sound bite." In my thinking, a very dangerous sound bite.

For about two and a half years, I was uninterested in a relationship with God. I allowed the self-proclaimed "Holy Warriors" to define my view of God, and I ceased to see Him as a force of love. God was, at best, a source of political manipulation—at worst, a source of pain, destruction, and death.

In the aftermath of Katrina, I returned to God, but the initial transformative rush was beginning to give way to disappointment and distrust, just as the day-to-day struggles of life—our relationships, our habits, our fears, our doubts, our shortcomings, our shame—can crush the spirit and separate us from God.

I composed myself there on the side of the road and called Shantrell, who was waiting for me to pick her up.

"How was it?" she asked.

"I'm so upset I cried. I don't cry much. Maybe we should have organized a protest. At least I might feel a little better."

"Protest? We're on a mission of mercy. When are you gettin' here?"

Shantrell was right. We, not our government, had a mission—a mission of mercy. My saving grace, so to speak, was that Katrina's survivors had shown me each day can be a new day through the love of God—always there, always faithful, but rarely what we expect.

eighteen

CHARITY

My thirteen-year-old goddaughter, Emma, told me the reaction of her classmates in Ann Arbor, Michigan, after hurricane Katrina: "Initial outrage," she said. Then: "George Bush is a horrible president. Then nothing. Forgotten."

The reactions of Emma's peers to Katrina could have been applied to many human crises across the globe—blame George Bush (or some other leader) and go on with life. But Katrina's survivors taught me that I could not make my disappointment in George Bush or our government, however legitimate, an excuse for my own inaction.

I learned a useful saying while training as a newly hired pilot at United Airlines: "It's not *who* is right, it's *what* is right." The training emphasizes that safety is the primary value—above rank, egos, and personalities. There is no excuse for failure to act properly to ensure safety, even when you are not in charge or when those in charge are jeopardizing safety. Katrina's survivors forced me to apply this standard to my life. Overwhelmed by images of desperation, I had to act; my world was going to crash if I did not try to take control and pull it out of its nose-dive.

While many people reacted as Emma's classmates did, dozens of others shared my feelings, and they tried to volunteer through the Red Cross and other organizations. Unfortunately, just as Katrina's survivors could not break through the stretched federal bureaucracies, volunteers could not break through the over-burdened national relief bureaucracies.

I spoke with a doctor in Florida, a bioenvironmental engineer in Michigan, a trauma-trained specialist in Maryland, and others who tried to go to the Gulf Coast but stopped trying after dozens of failed attempts over the phone or on the web. They wanted to go at their own expense, but their calls and e-mails were not returned by the relief organizations.

I aired the frustrations of would-be volunteers to several congressional staffers and asked if there was a nationally organized volunteer system that could link volunteers with those in need.

"Yes, I think there is a national system. I think it's called 2-1-1," one of them told me. There is a 2-1-1 helpline. More than thirty million Americans have access to it, according to the 2-1-1 website, but only nine states have fully functioning systems. Two 2-1-1 bills are pending, one in the Senate and one in the House. Senator Hillary Clinton (D-NY) and Representative Michael Bilirakis (R-FL) both sponsored the legislation in early 2005. Both bills are still sitting in committee, according to the Thomas.gov search engine.

Senator Clinton's bill states, "2-1-1 would connect individuals and families seeking services, volunteer opportunities, or both to appropriate human service agencies, including community-based and faith-based organizations and government agencies."

The bill also states, "There are approximately 1,500,000 nonprofit organizations in the United States. Individuals and families often find it difficult to navigate through a complex and ever-growing maze of human service agencies and programs, spending inordinate amounts of time trying to identify an agency or program that provides a service that may be immediately or urgently required and often abandoning the search from frustration or a lack of quality information."

Hillary Clinton's bill cites a University of Texas study that "esti-mates a net value to society of a national 2-1-1 system approaching $130 million in the first year alone and a conser-vative estimate of nearly $1.1 billion over 10 years."

Katrina survivors *needed* a 2-1-1 helpline. Instead, they mostly got old jeans and sweatshirts. The first sign of charitable giving I saw in the Gulf was an empty parking lot littered with old clothes. There were a few women picking through the piles of discarded clothing in the peak heat of the day. Their tired faces bore the look of sleepless nights and worry. I doubt they ever dreamed life would require them to shop in a 110-degree parking lot to clothe their children with old shirts and pants too worn or too tight for their original owners.

I began to question these makeshift thrift shops within about an hour of arriving in the Gulf. Looking back, my doubts began in New York. I was waiting for the elevator in my apartment building when I overheard a woman asking an elderly gen-tleman to give some of his things to the Katrina relief effort. Trucks were being loaded with supplies and clothes for the Gulf. "They are right across the street at Lincoln Center. They will take anything. Don't you even have an old shirt, or some-thing that doesn't fit?" the woman pleaded. The man shook his head and waved her off.

"Greedy bastard," I thought to myself. "He can't even give up a shirt." But after seeing the piles and piles of waste created by the donated clothes, I was glad he didn't turn loose of his old shirt. To a degree, however well-intentioned, these clothing drives didn't serve the greater needs of the survivors. In fact, the donated old clothes served those giving more than they served those in need. They enabled people to clean out their closets and say, "Yes, I gave to Katrina," and move on.

Katrina forced me to take another look at my own charity. Before the hurricane, I was just one level up from sending old clothes—I was pleased with myself for writing checks. I had bought into the business of giving. As my friend Bob Giuda, Chairman of Americans for Resolution of Kashmir, said on his way to Pakistan to build shelters for the earthquake victims, "Americans have made charity a business. We give a hundred dollars and go on about our lives. Charity is nothing more than a relief distribution mechanism."

Unfortunately, the typical relief distribution mechanisms were wholly inadequate in the Gulf Coast. The Red Cross is the largest relief distribution mechanism in the world. While its size and capabilities are impressive, it is not, nor is it meant to be, the alpha and omega of disaster relief. Nevertheless, the Red Cross provides a hassle-free giving mechanism. Payroll deductions and credit card contributions seem to serve us well. Some of us get to earn frequent flier miles and tax deductions while we're at it. While this relief distribution mechanism is very convenient for the giver and for those collecting the gifts, it does not fully serve the would-be receivers.

I asked several people in Mississippi what assistance they had received from the Red Cross. They told me they got ice chips, and a few got some money. They were grateful for both. Of course, the Red Cross has done more than sprinkle around ice chips and a few dollars. It operated many shelters far away from the devastated area, but the perception of people who remain on the coast is that the Red Cross has done little.

One reason for this seeming lack of aid is that the Red Cross's distribution system is centralized and inaccessible to many. "Don't talk to me about the Red Cross," one man told me. "You have to stand in line all day and night. For what? They say they may give you a few hundred dollars, but I work."

This man considers himself fortunate that the reason he can't stand in a Red Cross line or any other line is because he has a job. Others are less fortunate. They don't have jobs and they cannot stand in lines because they are too elderly or disabled.

I soon learned that the sources of greatest relief were volunteers who came on behalf of their neighborhoods, communities, and, most of all, their religious organizations.

One woman who spoke up during Bill Clinton's visit said, "The only people we have seen here are from the churches. And they come from churches from ALL over the country." The crowd clapped and cheered in affirmation.

In some cases, outside religious groups coordinated their efforts with the local churches; in other cases, they just showed up for work and went house to house. They were self-contained, autonomous agents of relief. They didn't have to answer to headquarters or to a legal team. There will always be a need for large relief organizations like the Red Cross. Not everyone can load up a U-Haul and drive to the nearest disaster. Nevertheless, Katrina forced me to question my own style of giving and the larger scope of modern charity. Before Katrina, I had kept my giving at arm's length—a check or payroll deduction, or perhaps a short talk about aviation to an audience of school-age children. I was helpful, but I was unwilling to enter into the lives of those in need. I was willing to go to the front lines of war, but not to the front lines of charity. As a result, I was missing out on one of life's most precious gifts—joy. Dr. Mitch Gaynor, my family physician, explained the joy of giving to me a few years ago during a physical right after the Christmas holidays. I asked him how his holidays had been and he said, "Great. We fed a town of people in India Christmas dinner." Then he asked, "Do you know the difference between happiness and joy?" He explained that happiness is short-lived. For example, we get something we want and we are happy for a little while, but it wears off. "We write a check to something we care about and we are happy for a moment. Joy, however, is lasting. Joy comes from not writing the check, but seeing the look on the face that receives the gift—being *part* of that gift."

Dr. Gaynor was explaining the same concept taught in the gospels of Christ. Jesus said that we find ourselves by giving ourselves away; that it is more blessed to give than to receive. But until Katrina, I had not touched the person who would receive my gift. In human contact, transformation began for me—and perhaps for them as well.

One Sunday Rev. Rosemary asked me to come and tell the congregation what we were doing for their community. I ended up telling them what they had done for me. I told them that I wanted to introduce America to them in hopes that their spirit and sense of community and love would heal our nation as it had healed me.

Compassion is the mother of charity. To be compassionate is to enter into that place of pain and suffering with another, even when we cannot fully heal their wound. It is in the act of being there—silent and in unity—that we find the strength to heal and be healed.

nineteen

GOD'S WILL

One afternoon Shantrell, Myrick, Ellen, Jason, Adam, Rev. Rosemary, Rev. Theodore, Ms. Leona (the trustee at Mt. Zion), and I loaded up in the church van and made our way past the military checkpoint that was designed to protect Pass Christian from its citizens (or its citizens from Pass Christian). The Pass is only a few minutes' drive from DeLisle. It's difficult to determine where DeLisle ends and the Pass begins. The Pass was more densely populated, but "dense" is a relative term. I imagine that the Pass was among the last of the small, affordable beach communities in America. The combination of wind and tidal surge damage was powerful enough to have moved homes fifty feet. We saw an abandoned U-Haul truck larger than ours suspended on top of a twenty-foot pile of debris, looking as if a crane had delicately balanced it there. We could see that several structures had not been searched for bodies. The numeral 0 was missing from the lower quadrant of the bright orange spray-painted X. That number would have indicated that the structure had been searched and no bodies had been found.

We pulled into the parking lot of St. Paul United Methodist Church. It looked fine from the outside. St. Paul is larger and more modern than Mt. Zion. It is half red brick and half wood siding. A large, long cross is mounted on the front, and a church bell sits on top of a concrete stand. The spray-painted X was on the church bell stand. It had the letters OH on the left quadrant, probably signifying that the Ohio National Guard had searched it, and the numeral 0 in the lower quadrant.

There appeared to be a message inside the glass case that held the pastor's name and the times of church services. We struggled to open the glass door, and then we struggled to read the water-soaked note. "Reverend Williams," it read. "Please call. We are okay." A name and a phone number followed. Rev. Theodore's face lit up to know that another one of his flock had survived. "Well, all right. Yes. I will call her. Well, all right." Several members of Rev. Theodore's congregation were still missing.

I asked Rev. Rosemary if Rev. Theodore had lost any members. She said yes and paused for a moment as if she was paying respect to the dead. "A young man. He stayed on through the storm." She tilted her head slightly downward and looked as though she was picturing him in her mind.

We opened the door; I was overcome by the smell of sewage and mold. The inside looked like the homes and apartment buildings I had visited. The mud and mold went from floor to ceiling. The heavy oak pews were still upright, but nothing else was recognizable. As I started taking video footage, Rev. Theodore said, "Be careful now. It's very slippery." The mold burned my lungs and the sewage made me want to gag. "Now don't stay too long. It's not safe," Rev. Rosemary continued.

By the time I got outside, I felt as though I had just been to a memorial service—a memorial for St. Paul United Methodist Church. I had never known her, but I felt an acute sense of loss through her pastor and her members—all of whom had lost their home.

We stood outside the church and tried to clean the toxic sludge from our shoes before getting back in the church van.

"Well look what blew in here!" Rev. Rosemary exclaimed as if she had just found a great treasure. She had found a bird-house made in the form of a little chapel. She picked it up to

examine it. Rev. Theodore smiled. "My goodness." Their sadness lifted, as if this tiny birdhouse was a sign of reassurance:
You are not forgotten.

From St. Paul we drove to the waterfront. There was nothing
left standing on the shore. Sand dunes were still piled high
along Highway 90. "I think that used to be a hotel," Shantrell
said. All that remained was the swimming pool.

"Look at that church!" All that remained of a church probably three times the size of St. Paul was its frame and bell.

It was the first time past the military checkpoint for some in
the group. After the initial exclamations of disbelief and shock,
we stood quietly. There were no words. The gentle breeze and
rhythmic sound of the Gulf waters yielded a strange sense of
peace in the midst of enormous loss. As we walked back to the
van, Ms. Leona, whom Ellen had nicknamed "the Church
Lady," broke the silence. "God's got a message in all this."

Ms. Leona gives off an air of moral authority—not holier-than-
thou, but You-better-stay-on-the-right-side-of-God. Personal
piety is nonnegotiable for Ms. Leona. As she said in Sunday
school one morning, "Every time we tell a little lie, a part of us
dies." She then looked around at the class with her eyebrows
raised, nodding her head until the rest of us nodded our heads
in agreement. I heartily agreed, based on experience as well as
my fear that she might come over and smack me. Ms. Leona is
shorter than I am, but looks like she could bench-press two or
three of me. She is a pillar of Mt. Zion. She is always at the
church—on guard and ready to help.

None of us responded to Ms. Leona's statement. She
repeated it. "God's got a message in all this."

"What do you think that message is?" I asked.

"You can't hide. I'm comin'. Better get your life straight," she
replied with certainty.

Ms. Leona was referring to the Second Coming of Christ (Revelation 3:3). Katrina was a sign of the Last Days as described in the gospels of Matthew 24:3–8, Mark 13:4–8, and Luke 21:7–11. I had heard this explanation for Katrina's destruction before.

Christians have been expecting the Second Coming of Christ for two thousand years. I am not one to speculate as to whether we are now living in the last days, but the truth is—on any given day, at any given time, in any given century—humans try to explain the unexplainable by blaming or crediting God. This is part of the greater phenomenon called God's will.

I became even more familiar with the term "God's will" while I was in the Middle East where the Arabic phrase *insha'Allah* ("if God wills") is repeated in nearly every conversation. The Qur'an instructs that no future planning be done without invoking this phrase. (Qur'an 18:23.)

The phrase *insha'Allah* is used when a person wishes to plan for the future, when he promises, when he makes resolutions, when he makes a pledge. For each, he asks the permission and the will of Allah. Muslims strive to put their trust in Allah and leave the results in Allah's hands.

The phrase is as common as "okay" in our conversations. I might say to a Muslim, "Okay, I will see you tomorrow." And the reply would be, "Yes, *insha'Allah*."

I learned of the term in 1993 while flying U-2 reconnaissance missions out of Saudi Arabia. The Saudi Ministry of Defense housed our small squadron in trailers. We nicknamed our new home "Camp Infidel." In accordance with this name, the Saudis decided that we needed a fence (a wall, actually), as their American guests were becoming something of a local tourist attraction.

You would think this fence was the Great Wall of China for

the time it took to construct it. The contractors were less than reliable and marginally skilled. Each time our commanding officer complained to them about the timeline or the crooked wall, they would smile and say, *"Insha'Allah."* We finally accepted them and their crooked wall, and dubbed them the Insha'Allah Brothers Construction Company.

While we joked about the Insha'Allah Brothers, the underlying philosophy that everything in life—even whether you show up at work on time or can lay bricks straight—hinges on God's will was difficult for me to accept.

I'm a recovering control freak and thrive on taking responsibility for my life and the world around me. I believed that it was up to me, not God, to get the job done. I thought that "let go and let God" was a copout, an excuse. If I interjected God into the picture, I was shirking my own responsibility. I had resisted the "God's will" attitude all my life.

Part of my resistance stemmed from the way the expression is generally used. When something bad that can't be explained happens, it's chalked up to God's will. "It's God's will that you lost your job." "It's God's will that you have cancer." "It's God's will that your infant child died." "It's God's will that your home and livelihood were destroyed." God's will is punitive.

Katrina let me to look beyond seeing God's will as a function of his controlling us as though we were puppets on the end of a string, and opened my eyes to seeing myself as an instrument of God's will, and God's will as an instrument of love, not punishment. My heart opened to the possibility that God's will is not concentrated on one terrible event, but that it is an ongoing, living, breathing expression of his love.

A sign hanging in the Harrison County emergency operations center says it best: "Hurricane Katrina was a force of nature. What we've done after it is an act of God."

I asked Ms. Leona if the message she thought God had for us in all this could be that no matter how bad things get, you still have each other, you still have your love for one another, and on that love you can build something greater than what was destroyed.

I finally understood my role in God's will. I became an agent of God's power, not a victim. I began to see that God's power is love, not judgment. Katrina's survivors let me be part of an "act of God."

One day while we took a break from putting up blue roof tarps, Rev. Rosemary told me how her vision for the community center and housing development had changed her. "I have had a comfortable life. We had things, you know. But this has shown that God can really use me." She seemed to be lost in her thoughts; perhaps she was contemplating the magnitude of the task before her.

God's call to service is akin to being chosen to fly a special mission. The difference is that this special mission is not one flight. It is not a goal. It is a daily opportunity and a dream. Every day I spent in Mississippi and each day since, I have looked at life through the lens of God's love.

My brother Chip once told me with absolute certainty that the purpose of life is to love God and the love of God is expressed through the love of one another. To paraphrase the apostle Paul (Romans 13:8–9), the greatest commandment is that we owe no one anything except to love one another, for the one who loves another has fulfilled the law. The commandments that we shall not commit adultery; shall not murder; shall not steal; shall not covet; and all the rest, are summed up in this one statement: love your neighbor as yourself. Through the expression of love, the act of giving, I regained my soul.

epilogue
SAVING THE SOUTH

In 2005, Mississippi was ranked last in an annual "Most Livable States" study for the seventh year in a row. According to the *Morgan Quitno Press*, the study is based on forty-four factors ranging from infant mortality rate to per capita income—which was $26,650 at the last U.S. Census Bureau estimate, and that was when times were good. It's no wonder the Mississippi state patron saint is Our Lady of Sorrows.

"I know Mississippi is ranked last in a lot of things. But there are a lot of us who are workin' hard to change things," said Ms. Rebecca Endt, who teaches world history and psychology and coaches volleyball and softball at Gautier High School. "Workin' hard" was an understatement. I had the opportunity to visit Gautier High School. Myrick Nicks and Anthony Herbert are the assistant vice principals. They, along with Principal Bernard Rogers, the rest of the administrative staff, and the teachers, are committed to educate and mentor the young adults who will ultimately lead Mississippi out of last place. Almost six months after the storm, they had not been given additional financial or material resources to cope with the added stress of the storm. They were making up the difference out of their own pockets.

When I met Principal Rogers, he had just come from buying some school uniforms for his students on his lunch break. Myrick and Shantrell also shopped in the evenings and weekends in order to clothe the needy students.

Myrick always kept a positive attitude, but I could see that he was worried about his students. Ellen asked Myrick one night as we were sitting around talking, "What do you do when you see your high school kids living in these horrible situations?" She was referring to pre- and post-Katrina. "Don't you just want to save every one of them?"

"Yes, I do. But after a while, you realize that as much as you want to, you can't really do that. The best thing we can do is to create a great environment for them while they are at school. We do the best we can in the time we have with them," Myrick replied.

His teachers and staff are equally committed. Ms. Anita Lawrence, who teaches special education, informed Anthony that one of her students was still sleeping on a wet mattress from the storm. His family had somehow been overlooked for assistance. Anthony asked the teacher to collect a list of needs and within a few hours, she had gone to the home and completed a list that was as all-encompassing as any I had seen. The family did not have pillows, blankets, towels, underwear, shoes, pants, or toilet paper. They only had the love of a teacher who valued their child as though he was her own.

Shantrell and I delivered the supplies to Ms. Lawrence's classroom after I returned from President Bush's visit. I admired her courage. I cannot imagine what it would be like to have to look into the eyes of those students every day and know that they lack the most basic of human needs.

Some of the teachers I met were sharing their land so that other staff members would have a place to put their tents or trailers.

I watched as Anthony and Myrick patrolled the halls between classes. "This is a quiet zone, Gators. Quiet in the halls." Shantrell joked that it was like boot camp. Personal responsibility and respect for oneself and others were the foundations of learning and teaching.

A large sign in the school summarized Gautier High School's values. In the middle of the sign was the word EXCUSES circled in red with a red line through it. Underneath it said, "This teacher makes none nor accepts any." These quotations were printed in each of the four corners of the sign:

"Ninety-nine percent of failures come from people who have a habit of making excuses." —George Washington Carver

"You've got to continue to grow or you're just like last night's cornbread—stale and dry." —Loretta Lynn

"Heart is what separates the good from the great." —Michael Jordan

"One important key to success is self-confidence. One important key to self-confidence is preparation." —Arthur Ashe

No one would know by looking at these students, teachers, or staff members that many of them no longer had homes, that they were taking their only shower for the day in the school gym, or that they had no money to repair their homes. They smiled and joked about their situations and spoke of their community with love and pride. They were focused on the future.

Their faith in a better future reminded me of a man I met the first night we arrived in Gulfport. Mr. Harry McInnis came over to take a group picture for us after we had unloaded the relief supplies the first night we arrived in Gulfport. "I'll make sure you get a copy," he said. "I can make a print. I can put it on a disc. Anything you want."

I thanked him and asked him what he did for a living. "Well, I'm a photographer. I'm a mechanic. I'm a preacher. I'm anything I need to be. I'm lucky. My garage got wiped out by Katrina, but I found another job."

"So you're a renaissance man," I said. He smiled and changed the subject, "You know, my son Jarvis is going to be the first black president of the United States."

He then launched into a talking tutorial on the life and achievements of Jarvis. "Right now—right now—he's in Washington, D.C., competing to win National Youth for the whole nation. He's going to win it, you know. It's all over but the celebrating."

"What's National Youth?" I asked him. He looked at me as if I might be from another planet. "You know. It's National Youth . . . for the entire nation."

Our paths crossed a few days later. We were gutting the home of Harry's neighbor, Ms. Verlon. Harry came over to see what we were up to and asked if I would take a look at his home. He was worried about it.

"Did Jarvis win National Youth?" I asked.

"Well, of course he's won. I mean—he won before he ever got there. We're just waiting for them to announce it."

I walked across the street with him. Katrina had taken her toll on the McInnis home, but Harry had done a remarkable job of making it look as if nothing had happened, with the exception of the ubiquitous blue tarp nailed to the roof.

Harry is a talker. He told me about every repair he had made after the storm. "Can you smell any mold?" he asked. I didn't want to break the news to him that much of his home would need to be stripped, just like Ms. Verlon's home. I tried to frame the news positively. "Your home smells wonderful because of those candles your wife is burning. But I'm afraid you do have some trouble spots." As I began to point out the areas of concern, I came across "The Wall of Jarvis."

A two-page article titled "When Jarvis Speaks, People Listen," was tacked to the wall. The picture with the article was taken from behind Jarvis as he was addressing a huge crowd. He was standing tall, like his father. The article was mounted on wood, then coated with polyurethane. It was preserved for eternity.

As we walked back over to Ms. Verlon's, Harry told me he had to leave to prepare to preach Sunday's service, and then he was going to drive to Jackson, Mississippi, to deliver Jarvis's first college term paper. Jackson is two hours away. When I protested, saying that was a long way to have to travel to deliver a paper under these circumstances, Harry simply said, "It's due."

That evening as I was relating the events of the day to Shantrell, Myrick, Jason, Adam, and Ellen. I told them about the Wall of Jarvis. Shantrell smiled and said, "I read that article. Jarvis really is an amazing speaker. The remarkable thing is that he speaks with a stutter."

A few days after we had left Mississippi Shantrell called and said two words: "Jarvis won!" Just as his father so confidently predicted, Jarvis McInnis was named "The Boys and Girls Club of America 2005–2006 National Youth of the Year." Jarvis won as a member of Mr. Purvis McBride's Boys and Girls Club in the Turnkey development where we'd helped out. According to the press release, Jarvis will receive a $15,000 scholarship from the Reader's Digest Foundation and may meet President Bush.

Here is an excerpt from Jarvis's award-winning speech:

> We've often heard the popular cliché, "Sticks and stones may break my bones but words will never hurt me." Well, words do hurt—and they hurt me. I have a speech impediment. I stutter and have been doing so all my life. While at school, my peers would mock me, taunt me, and laugh at my struggle to pronounce certain words. Even some of my family, my own flesh and blood, made fun of me. Though I tried to ignore their harsh words, I just couldn't do it . . . and the tears of frustration would fill up in the wells of my eyes. And I would just cry.

But my Boys and Girls Club was there for me. I am no longer taunted by those harsh words. My club built my self-esteem and focused on my strengths. My club gave me a sense of belonging—a sense of competence and usefulness by teaching me to turn my pain and frustrations into positive, creative energies.

Harry McInnis will need to make room on the Wall of Jarvis for a new award, and our nation will need to make room for a new leader.

When Shantrell called us with the good news, Ellen and I cried the tears of joy reserved for those moments when justice, faith and hope prevail in a world of pain and loss. Katrina has washed away lives, homes, businesses and jobs, but she could not wash away a father's confidence in his son or a community's hope for a better future.

Hurricane Katrina is the worst natural disaster in American history. As Myrick once said, "It's like we were a bunch of ants on our ant hill living life, doing what ants do and then all of a sudden someone comes along and kicks our hill way up in the air and destroys our whole world."

Myrick, Shantrell, Rev. Rosemary, Rev. Theodore, Anthony, Sonya, and the hundreds of other people I have met in the Gulf Coast understand that if they do not take responsibility for rebuilding, it will not be rebuilt. Their community will be lost. They will continue to do their part. But they also understand that there will be no community, no businesses, and no jobs without homes. Homes cannot be rebuilt or repaired without money in the form of loans and/or grants.

Our commitment to Katrina's survivors must outlive the length of time she remains in the headlines or we will create a social crisis of homelessness and unemployment that will last for decades. Individually and collectively, we will reap what we sow.

The people of the Gulf Coast of Mississippi and those who have come to their aid have shown me that the human spirit, powered by the force of love, soars above bureaucracy, neglect, and injustice. Katrina transformed my view of life, forcing me to view destruction and loss through the lens of love. It is through the eye of the storm that I can clearly see the power of God's love manifested in the charity, service, and sacrifice of the weak, weary, and worn. The storm has cleared a path for hope that can only survive if we are faithful to our fellow brothers and sisters in their hour of need. Christ said in Matthew 7:24, "Everyone then who hears these words of mine and acts on them will be like a wise man who built his house on rock. The rain fell, the floods came, and the winds blew and beat on that house, but it did not fall, because it had been founded on rock."

ACKNOWLEDGMENTS

No words could adequately express my gratitude to the people of the Gulf Coast who inspired me to write this book. My gratitude will have to be measured by my actions and not my words as we collectively work to build a new future for them and their children.

Of course this book would not have been possible without Margo Baldwin and the Chelsea Green Publishing Company. They felt strongly enough about Katrina's survivors to add this book to their spring catalog at the last minute. They took a risk with a first-time author who also had a full-time job and a two-month deadline. Their love for their work, their world, and one another is an inspiration. This was a team effort. The editor-in-chief, John Barstow, has the patience of a saint, the mind of a maestro, and a heart of gold. His insight, guidance, and brilliance have made all the difference in this book. Collette Leonard turned around the final copy edit practically overnight despite having a newborn at home. Marcy Brant and Jonathan Teller-Elsberg also worked odd hours and overtime in order to publish the book as soon as possible.

Fortunately for me and for Chelsea Green, my mother, Sharolyn Johnson, edited the book first. I had not been humbled by my mother's red pen for twenty-five years and had forgotten how talented she is. In addition to sentence structure, grammar, and clarification, she added perspective and balance that would have been difficult to find without her. Her work ethic and integrity have provided the example for me in my life and writing.

My brother, Chip Espinoza, has given me a much wider-angle lens from which to view the world, relationships, and faith. This book is full of his wisdom and guidance. He has always been and will always be my big brother and my hero.

I am grateful to my colleagues at Talk Radio News Service in Washington, D.C., Lovisa Frost, Greg Gorman, Victoria Jones and Wendy Wang. We share a common profession and values. They have diligently tracked government relief efforts in the Gulf through congressional hearings, research, and interviews.

Tami LeHouillier, Alexandra Marks, Richard Miller, and Richard Siklos are friends and mentors. They generously gave their time and constructive criticism. Their guidance and encouragement is reflected in this work.

There would not have been a book without my partner, Ellen Ratner. It was Ellen's idea to help Mt. Zion. It was Ellen's idea to write this book. I am blessed to share the gift of life with someone who believes with her every fiber that life is indeed a gift—a gift to be cherished—a gift to be shared.

Ellen's family has also shared in our commitment to the Gulf Coast. Her brothers, Bruce and Michael, generously financed our initial relief efforts, and her cousin, Ronald Ratner, has provided expertise in order to begin building a community center for Pass Christian and DeLisle, Mississippi.

As they say, "success has a thousand fathers" and *mothers* in this case. This book embodies that truth.

KATRINA BY THE NUMBERS

Number of housing units damaged, destroyed, or inaccessible because of Katrina: 850,791

Number of churches, synagogues, and mosques damaged or destroyed: approximately 900

Number of homes destroyed by breaches in federally designed and funded levees and not covered under the federal housing recovery plan: 200,000

Amount committed to Katrina relief by the federal government: $85 billion

Spent by FEMA specifically on housing assistance for hurricane victims: less than $4 billion

Spent by FEMA on operating expenses, including salaries and expense accounts: $6 billion

Spent on administrative overhead for every dollar FEMA spends: 26 cents

Number of FEMA trailers occupied in Mississippi: 94,000

FEMA trailers still needed in Mississippi: 9,000

FEMA trailers requested in the New Orleans metro area: 69,706

FEMA trailers occupied in the New Orleans metro area: 31,517

Unoccupied modular homes purchased by FEMA and sinking into mud in Hope, Arkansas: 10,777

FEMA trailers held in staging areas and unoccupied: 20,000

Repair and maintenance requests for FEMA trailers in Mississippi: 34,000

Average cost of a single FEMA trailer per month: $3,200

Cost to taxpayers for debris removal per cubic yard: $32

Payment to subcontractors for debris removal per cubic yard: $6-10

Number of "evacuees" given FEMA emergency assistance with invalid Social Security numbers or false addresses and names: 900,000

Percentage of FEMA contracts that were "no bid" in September 2005: 80

In October 2005: 60

In November 2005: 68

In the first half of December 2005: 50

Percentage of FEMA contracts by mid-November 2005 that went to firms in Alabama, Louisiana, and Mississippi: 12

Number of new migrant workers to the Gulf Coast region since Katrina: 30,000

Percentage of New Orleans' pre-Katrina residents who have returned to the city: approximately 40

Percentage of homeowner settlements with insurance companies by January 2005 after the four Florida hurricanes: 90

Percentage of homeowner settlements by February 2006 after Katrina: 70

Average homeowner claim for flood damage before Katrina: $22,084

After Katrina: $93,118

Number of insurance companies instructed by FEMA to cease National Flood Insurance payouts due to insolvency of the federally managed National Flood Insurance Program: 96

Amount allocated from Katrina funding to date to pay National Flood Insurance Program claims: $18.5 billion

Number of insurance companies sued for refusal to pay damages: 50

Number of counts in Senator Trent Lott's lawsuit against State Farm Insurance: 7

Insurance industry's contributions to Democratic campaigns and PACs for the 2004 and 2006 election cycles combined: $15,101,286

Insurance industry's contributions to Republican campaigns and PACs for the 2004 and 2006 election cycles combined: $31,282,859

Percentage of homeowners still awaiting Small Business Association disaster loan approval: 50

Percentage of homeowner SBA disaster loans that have been fully paid after approval: 6.9

Amount collected by The American Red Cross' hurricane relief fund: $2.1 billion

Annual salary of former Red Cross CEO Marsh Evans: $651,957

Amount paid to consultants in the past three years to boost the American Red Cross's profile: $500,000

Gallons of crude oil contaminating 2,500 Louisiana homes: 1,000,000

Number of medical professionals who volunteered with the Department of Health and Human Services after Katrina: 30,000

Number called to serve: approximately 1,400

Number of Katrina victims still missing: 1,960

Number of missing victims 20 years old or younger: 245

Sources (October 5, 2005–February 27, 2006):
1, 2, 30, 32, 41, 44, 45 *USA Today*. 3, 7 Website of Senator Mary Landrieu (D-LA), http://landrieu.senate.gov. 4 Speech by President Bush in Waveland, Mississippi, on January 12, 2006. 5, 6 White House Office of Management and Budget, provided to Eyewitness News WWL-TV. 8, 9 *Gulf Coast News*. 10, 11, 17, 25, 36 *The Times-Picayune*. 12 DHS audit, February 2006. 13, 15 *American Chronicle*. 14, 40 *The Washington Post*. 16 Senator Tom Coburn. 18 GAO Report. 19, 20, 21, 22 www.federaltimes.com. 23 Gulf Coast Reconstruction Watch/Institute for Southern Studies. 24 www.migrationinformation.org. 26, 27 Newhouse News Service. 28, 29 www.fema.gov. 31 White House OMB and *Capitol City Press*. 33 *Trent and Tricia Lott vs. US State Farm Fire and Casualty Company and John Does 1-10*. 34, 35 www.opensecrets.org. 37 Senate Resolution 347. 38 The Associated Press. 39 www.forbes.com (FY 6/30/03). 42, 43 Department of Health and Human Services, quoted by www.reconstructionwatch.org.

HOW TO HELP

All of the author's proceeds from this book will go to the **Pass Christian/DeLisle Community Center, Inc.** The directors of the center include Rev. Rosemary Williams, Shantrell Nicks, Myrick Nicks, Ellen Ratner and Cholene Espinoza.

You may make a tax-deductible contribution to:

Pass Christian/DeLisle Community Center, Inc.
Attn. Shantrell Nicks
2108 23rd Avenue
Gulfport, MS 39501
228-864-8846
shnicksesq@aol.com

If you would like to contribute general relief funds, sponsor a child or family, or physically volunteer to help the communities of DeLisle, Pass Christian, and Gulfport, Mississippi, please contact:

Rev. Rosemary Williams
P.O. Box 657
Pass Christian, MS 39571
228-864-4299
rwms101@aol.com

You may specify how you would like your dollars to be spent: on housing, transportation, building materials, food, clothing, electricity, general Katrina relief, etc. Mt. Zion United Methodist Church has a 501(c)3 status as well.

Updates on the community center and recovery efforts as well as new volunteer opportunities can be found at www.throughtheeyeofthestorm.com. The website also includes audio files and video footage of the people described in *Through the Eye of the Storm*.

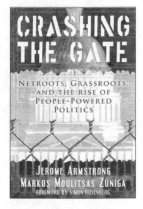
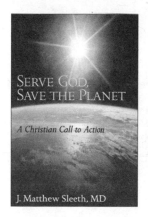